Collaborative Design in Virtual Environments

International Series on
INTELLIGENT SYSTEMS, CONTROL, AND AUTOMATION:
SCIENCE AND ENGINEERING

VOLUME 48



For other titles published in this series, go to
www.springer.com/series/6259

Xiangyu Wang • Jerry Jen-Hung Tsai (Eds.)

# Collaborative Design in Virtual Environments

Xiangyu Wang
Senior Lecturer,
Faculty of Built Environment,
The University of New South Wales,
Australia

and

International Scholar,
Department of Housing and
Interior Design,
Kyung Hee University, Korea

Jerry Jen-Hung Tsai
Yuan Ze University
Yuan-Tung Road 135
32003 Chung-Li
Taiwan R.O.C.
E-mail: jerry.tsai@saturn.yzu.edu.tw

ISBN 978-94-007-3514-9
ISBN 978-94-007-0605-7 (eBook)
DOI 10.1007/978-94-007-0605-7
Springer Dordrecht Heidelberg London New York



*Typesetting & Cover design:* Scientific Publishing Services Pvt. Ltd., Chennai, India

Printed on acid-free paper

Springer is part of Springer Science+Business Media (www.springer.com)

# Preface

Collaborative Virtual Environments (CVEs) are multi-user virtual realities that actively support communication, collaboration, and coordination. All of the academic books in this area are more focused on the theory, user-centred design, realisation and evaluation of Collaborative Virtual Environments in general. In contrast, the emphasis on studying designers' behaviours/actions/patterns in CVEs distinguishes this book from many general books which more deal with the design and development of CVEs. As more researchers in design and related areas progressively employ CVEs as their base of enquiry, we see a need for a reference guide bringing the existing status of CVEs into awareness and expanding on recent research.

This book offers a comprehensive reference volume to the state-of-the-art in the area of design studies in CVEs. This book is an excellent mix of over 16 leading researcher/experts in multiple disciplines from academia and industry. All authors are experts and/or top researchers in their respective areas and each of the chapters has been rigorously reviewed for intellectual content by the editorial team to ensure a high quality. This book provides up-to-date insight into the current research topics in this field as well as the latest technological advancements and the best working examples. Many of these results and ideas are also applicable to other areas such as CVE for design education. Predominantly, the chapters introduce most recent research projects on theories, applications and solutions of CVEs for design purpose. More specifically, the central focus is on the manner in which they can be applied to influence practices in design and design related industries.

Overall, this book offers an excellent reference for the postgraduate students, the researchers and the practitioners who need a comprehensive approach to study the design behaviours in CVEs. This book is a useful and informative source of materials for those interested in learning more on using/developing CVEs to support design and design collaboration. The target audiences of the book are practitioners, academics, researchers, and graduate students at universities, and industrial research that work with CVEs and digital media in a wide range of design areas.

The book has 5 sections and 16 chapters totally. The sections are listed as follows and more information can be found in the Table of Contents;

Part III. Design Cooperation: Sharing Context in Collaborative Virtual Environments
Part IV. How Designers Design in Collaborative Virtual Environments
Part V. Case Studies

## Part I. Virtual Environments for Design: Fundamentals

To begin, Professor Mary Lou Maher overviews the technical and social issues of CVEs and their impact on designers in her keynote chapter *Designers and Collaborative Virtual Environments*. This overview of CVEs sets the context and frames the scope of the book. It discusses how CVEs has lead to new ways for designers to collaborate and new kinds of places for designers to design.

Apparently, designing in virtual environments unavoidably involves visual-spatial cognition. The second chapter *Visual-Spatial Learning and Training in Collaborative Design in Virtual Environments* by Maria kozhevnikov and Andre Garcia reviews different types of virtual environments and discusses the major advantages that these environments can offer in the domain of visual-spatial performance. The first part is then followed by the following four parts in which there are chapters relating to more specific aspects of collaborative design in virtual environments.

## Part II. Representation and Embodiments in Collaborative Virtual Environments: Objects, Users, and Presence

This part highlights issues with the representation of objects and embodiments of users by avatars in CVEs. This part develops an understanding of the nature of presence in CVEs from real-world investigation of the means by which users experience CVEs. The three chapters in the second part present current research in this area.

Chiu-Shui Chan explored in *Design Representation and Perception in Virtual Environments* two important cognitive activities involved in designing in virtual environments. The first activity is design representation that is mentally created during the design processes. The second activity relates to human perception, which has not been changed by high-tech developments.

Form and content are two basic concepts that have a significant impact on the sense of presence in virtual environments. The second chapter by Rivka Oxman *Design Paradigms for the Enhancement of Presence in Virtual Environments* discusses current research in the design of presence in virtual environments.

Co-presence has been considered as a very critical factor in CVEs. Xiangyu Wang and Rui Wang follow Rivka Oxman's discussion in the third chapter, *Co-presence in Mixed Reality-Mediated Collaborative Design Space,* and reflect on the concept and characteristics of co-presence, by considering how Mixed Reality-mediated collaborative virtual environments could be specified, and therefore to provide distributed designers with a more effective design environment that improves the sense of "being together".

## Part III. Design Cooperation: Sharing Context in Collaborative Virtual Environments

The third part contains chapters addressing collaboration, communication, and coordination methods and issues in using CVEs for collaborative design activities.

The first chapter by Walid Tizani, *Collaborative Design in Virtual Environments at Conceptual Stage,* outlines the requirements of collaborative virtual systems and proposes methodologies for the issues of concurrency and the management of processes.

The second chapter by Jeff WT Kan, Jerry J-H Tsai and Xiangyu Wang, *"Scales" Affecting Design Communication in Collaborative Virtual Environments,* explores the impacts of large and small scales of designed objects towards the communication in three-dimensional collaborative virtual environments.

As a means for design coordination and progress monitoring during the construction phase, the chapter by Feniosky Peña-Mora, Mani Golparvar-Fard, Zeeshan Aziz and Seungjun Roh, *Design Coordination and Progress Monitoring During the Construction Phase,* presents a complementary 3D walkthrough environment which provides users with an intuitive understanding of the construction progress using advanced computer visualization and colour and pattern coding techniques to compare the as-planned with the as-built construction progress. The innovation of this method is to superimpose 3D Building Information Models (BIM) over construction photographs.

## Part IV. How Designers Design in Collaborative Virtual Environments

The fourth section looks at how designers design in CVEs. Nobuyoshi Yabuki sets out in his chapter, *Impact of Collaborative Virtual Environments on Design Process,* to review the current design and engineering processes and identifies issues and problems in design and construction of civil and built environments. Based on these findings, he then investigates and summarizes the significant impacts of CVEs on design and construction of civil and built environments.

As an effort to study how designers learn design in CVEs, in their chapter *A Pedagogical Approach to Exploring Place and Interaction Design in Collaborative Virtual Environments*, Ning Gu and Kathryn Merrick report on their experience of teaching the design of virtual worlds as a design subject, and discusses the principles for designing interactive virtual worlds.

Ellen Yi-Luen Do, wrote the third chapter, *Sketch that Scene for Me and Meet Me in Cyberspace.* It discusses several interesting projects using sketching as an interface to create or interact in the 3D virtual environments.

In the final chapter in this part, *A Hybrid Direct Visual Editing Method for Architectural Massing Study in Virtual Environments*, Jian Chen presents a hybrid environment to investigate the use of a table-prop and physics-based manipulation, for quick and rough object creation and manipulation in three-dimensional (3D) virtual environments.

**Part V. Case Studies**

This part collects 4 chapters on emerging technology implementation and applications of virtual environments in collaborative design.

Firstly, Bharat Dave, in his chapter, *Spaces of Design Collaboration*, emphasizes the socially and spatially situated nature of collaborative design activities and settings, and identifies issues that remain critical for future collaborative virtual environments.

In the following chapter, *Modeling of Buildings for Collaborative Design in a Virtual Environment*, Aizhu Ren and Fangqin Tang, present an application independent modeling system, which enables quick modeling of irregular and complicated building structures adapted to Virtual Reality applications based on Web.

Phillip S Dunston, Laura L Arns, James D Mcglothlin, Gregory C Lasker and Adam G Kushner present in their chapter, *An Immersive Virtual Reality Mock-up for Design Review of Hospital Patient Rooms,* the utilization of Virtual Reality mock-ups to offer healthcare facility stakeholders the opportunity to comprehend proposed designs clearly during the planning and design phases, thus enabling the greatest influence on design decision making.

In the final chapter, Marc Aurel Schnabel discusses in *the Immersive Virtual Environment Design Studio,* the implications of designing, perception, comprehension, communication and collaboration within the framework of an 'Immersive Virtual Environment Design Studio'.

**Acknowledgements**

We express our gratitude to all authors for their enthusiasm to contribute their research as published here. This book would not have been possible without the constructive comments and advice from Professor John Gero, from the Krasnow Institute for Advanced Study.

Xiangyu Wang
Jerry Jen-Hung Tsai

# Contents

# Part I
# Virtual Environments for Design: Fundamentals

Designers and Collaborative Virtual Environments
Mary Lou Maher (University of Sydney)

Visual-Spatial Learning and Training in Collaborative Design in Virtual Environments
Maria Kozhevnikov and Andre Garcia
(Harvard Medical School and George Mason University)

# Designers and Collaborative Virtual Environments

Mary Lou Maher

University of Sydney, Australia

**Abstract.** This chapter provides an overview of the technical and social issues of CVEs and their impact on designers. The development of CVEs has lead to new ways for designers to collaborate and new kinds of places for designers to design. As a new technology for collaborative design, CVEs impact the collaborative process by facilitating movement between working together and working individually. As new technologies for interacting with CVEs include tangible interfaces, we can expect to see changes in the perception of the design that lead to changes in spatial focus.

**Keywords:** 3D Virtual Worlds. Collaborative design, tangible interaction, protocol studies, adaptive agents.

## 1 Introduction

Collaborative Virtual Environments (CVEs) are virtual worlds shared by participants across a computer network. There are many descriptions of CVEs, and Benford et al (2001) provides one that is assumed in this chapter: The virtual world is typically presented as a 3D place-like environment in which participants are provided with graphical embodiments called avatars that convey their identity, presence, location, and activities to others. CVEs vary in their representational richness from 3D virtual reality to 2D images to text-based environments. The participants are able to use their avatars to interact with and sometimes create the contents of the world, and to communicate with one another using different media including audio, video, gestures, and text. This kind of virtual environment provides opportunities for collaborative design that gives remote participants a sense of a shared place and presence while they collaborate.

CVEs provide new ways to meet communication needs when negotiation is important and frequent, and complex topics are being discussed. They provide more effective alternatives to video conferencing and teleconferencing because they provide spaces that explicitly include both data representations and users (Churchill et. al. 2001), an important consideration in collaborative design where the focus of a meeting may be on the design ideas and models more than on the faces of the collaborating designers. During the early days of CVEs (in the early 1990s), researchers put an emphasis on simulating face-to-face co-presence as realistically as possible (Redfern and Naughton, 2002). More recently, it has been realised that

X. Wang & J.J.-H. Tsai (Eds.): Collaborative Design in Virtual Environments, ISCA 48, pp. 3–15.
springerlink.com 

this is not enough for collaborative design, and may not necessarily even be required to develop a shared understanding of the design problem and potential solutions (Saad and Maher, 1996).

Redfern and Naughton (2002) nicely summarize a range of technical and social issues provided in the development and use of CVEs in distance education that can be adapted for understanding the development and use of CVEs in collaborative design.

*Managing collaborative design processes.* In a design project, collaborative work involves the interleaving of individual and group activities. Managing this process over the several months of a design project requires considerable explicit and implicit communication between collaborators. Individuals need to negotiate shared understandings of design goals, of design decomposition and resource allocation, and of progress on specific tasks. It is important that collaborators know what is currently being done and what has been done in context of the goals. In a collaborative design task this information can be communicated in the objects within the CVE where the collection of objects forms an information model, such as the Building Information Model in building design processes. DesignWorld is an example of a research project that explores ways of integrating CVEs with an external database of objects and project information (Maher et al 2006).

*"What You See Is What I See"* (WYSIWIS). Conversational and action analysis studies of traditional collaborative work have shown the importance of being able to understand the viewpoints, focuses of attention, and actions of collaborators. CVEs assume a co-presence in a virtual world that is shared, even though the viewpoint of the world may be different when the avatars are located in different places and facing different directions. Communication among the participants in a CVE is often about location and viewpoints, allowing individuals to pursue their own tasks as well as have their attention focussed on a shared task. Clark and Maher (2006) studied communication in a design studio course that was held in a CVE and showed that a significant percentage of the communication was about location and presence.

*Chance meetings.* Informal meetings with colleagues are rarely provided for in collaborative tools, yet they are an important characteristic of the effectiveness of many workers, particularly knowledge-workers. Recent research has investigated mechanisms for supporting chance meetings without the requirement for explicit action by the user (McGrath & Prinz, 2001). In collaborative design, studies have shown that designers move fluidly from working individually to working together. Kvan (2000) presents a model in which different stages of collaborative design are characterized as closely coupled or loosely coupled. CVEs provide the opportunity for individual work in a shared place that supports chance meetings.

*Peripheral awareness* is increasingly seen as an important concept in collaborative work, as evidenced in ethnographic studies. Team members involved in parallel but independent ongoing activities need to be able to co-ordinate and inform their activities through background or peripheral awareness of one another's activities. The affordance of peripheral awareness for collaborative design in a CVE is demonstrated in a study done by Gul and Maher (2009). In this study, designers were given similar design tasks in a 3D CVE and in a remote sketching environment, and asked to collaborate for a fixed period of time. An analysis of the protocol data shows that in a 3D CVE designers were inclined to spend part of the time working together and part coordinating their individual work, while in a remote sketching environment the designers did not work individually.

*Non-verbal communications* are known to have a strong effect on how utterances are interpreted. Research into alternative input mechanisms for capturing this type of information from the user has been underway for some time: recently, attempts are being made to make these mechanisms intuitive and non-intrusive. Clark and Maher (2006) show how participants communicated using specified gestures for their avatars in the design studio. Augmented reality approaches to CVEs promise a more spontaneous integration of movement in the physical world being communicated in a virtual world.

The *"designing for two worlds"* principle: People are never fully immersed in a virtual world, but are always partially in the real world too. Certain activities when carried out in the real world have a very strong impact on the participant's activities that should be recognised in the virtual world – for example, answering the phone. DesignWorld (Maher et al 2006) accommodated this by maintaining a video of each designer in his physical office in a window adjacent to the 3D CVE with the designers' avatars. This allows communication to be directed in the virtual world or in the physical world, and the physical presence and activities of the physical world to be communicated to the designers in the virtual world.

This chapter provides an overview of two comparative studies of collaborating designers using CVE technologies. These studies provide a starting point for understanding the impact of these technologies on design cognition and design collaboration. The chapter ends with an overview of a project that considers the opportunities that CVEs provide for designers to explore a new kind of design discipline: the design of places in virtual worlds. These three projects consider designers more comprehensively in the context of CVEs: from designers as users of CVEs to designers of CVEs.

## 2 Supporting Collaborative Design: From Sketching to CVEs

Sharing design ideas ranges from working together at a table while sketching with paper and pencil, to working in a CVE. CVEs do not replace sketching on paper while co-located; they provide a different kind of environment for collaborating. Since the tools for expressing and sharing ideas are so different, we would expect that the collaboration is different. Gul and Maher (2009) describe a study comparing design collaboration while designers sit together sketching to remote sketching and designing in a 3D CVE. The aim of the study is to identify the changes in collaborating designers' behaviour and processes when they move from co-located sketching to remote designing.

The study considered three collaborative design settings: sketching around a table, remote sketching, and designing in a CVE. The sketching setting is shown in Table 1. The image in the left part of the table shows a camera image of two designers sharing a physical table with sketching tools such as paper and pencil.

**Table 1.** Sketching experimental setting (Gul and Maher, 2009)

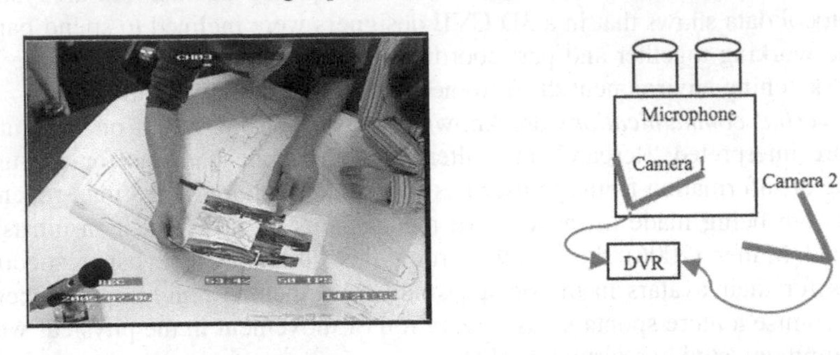


**Table 2.** Remote sketching setting (Gul and Maher, 2009)

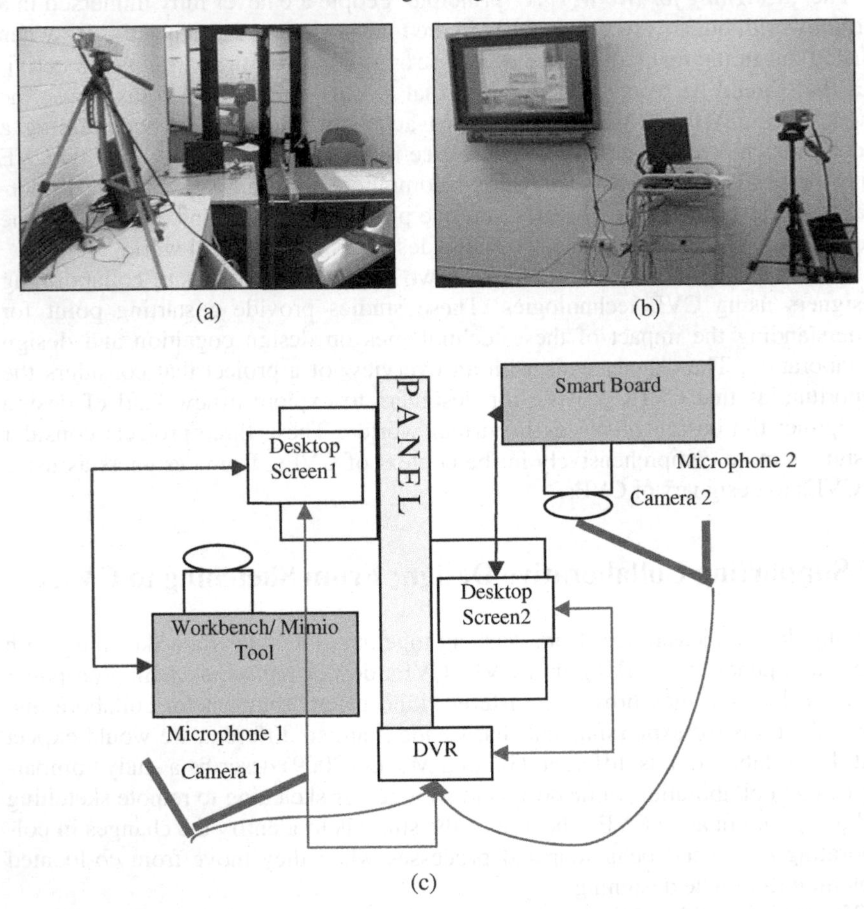


The schematic in the right part of the table shows how the designers were observed by placing 2 cameras connected to a digital video recorder.

The remote sketching setting used GroupBoard (www.groupboard.com), a collaborative sketching application, and display surfaces with pen interfaces. One

**Table 3.** Remote sketching and CVE interfaces (Gul and Maher, 2009)

(a) GroupBoard Interface

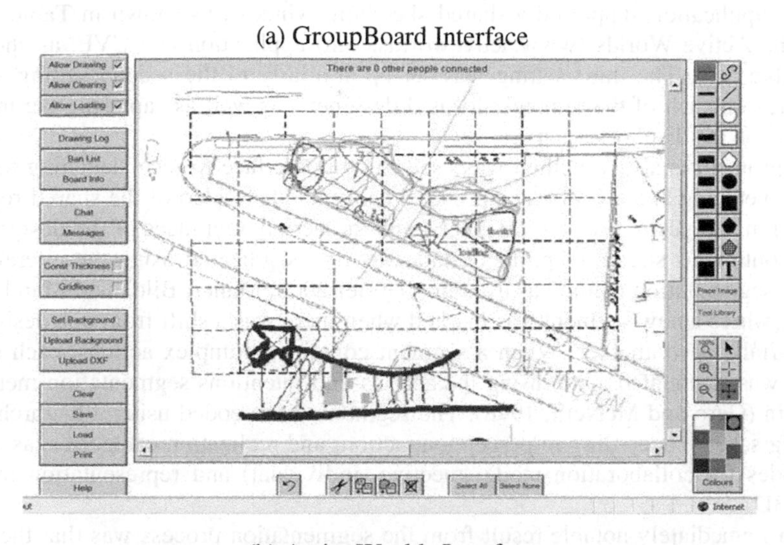


(b) ActiveWorlds Interface

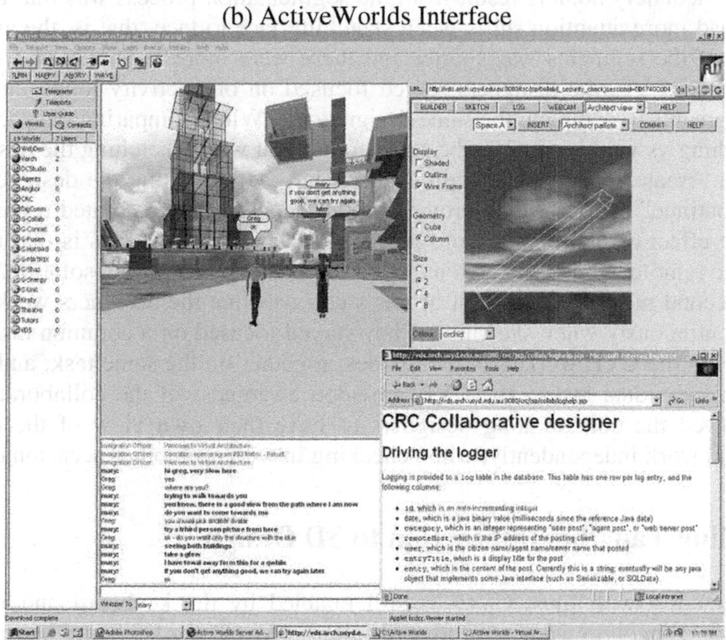


designer was sketching on a tabletop system and the other designer was sketching on a whiteboard mounted on the wall. The setting is shown in Table 2. The top row of the table shows the tabletop sketching environment (left) and the mounted whiteboard sketching environment (right). The bottom row of the table shows a schematic of the layout: the location of cameras for recording the design session, and the use of a panel to simulate remotely located designers.

Table 3 shows the applications for the two remote design sessions: the GroupBoard application supported a shared sketching window, as shown in Table 3(a) and the Active Worlds (www.activeworlds.com) application is a CVE, as shown in Table 3(b). The third setting was set up similarly to the second setting with cameras on each of the remotely located designers, as well as capturing the interactions on the display screen.

A protocol analysis method was used to compare face to face sketching to remote sketching and a CVE. The protocol data included video of the shared representation and gestures of the designers, and the verbal utterances of the designers. The continuous stream of protocol data was first segmented using the utterance-based segmentation method as used in (Gabriel 2000; Maher, Bilda and Marchant, 2005) where a new segment was marked when there was a shift from one designer acting/talking to another. When a segment contained complex actions, each segment was segmented again using the actions-and-intentions segmentation method used in (Gero and McNeill, 1998). The segments were coded using a hierarchical coding scheme according to perception, action, and realization categories, as well as codes for collaboration mode (meeting, individual) and representation mode (2D, 3D).

An immediately notable result from the segmentation process was that the designers had more attention shifts when sketching face to face, that is, the average duration of the segments was shorter and there were more segments in a design session. In a CVE, the designers stayed focused on one activity longer and had fewer segments to complete the same design goals. When comparing the protocols for sketching vs working in 3D, the study found that when sketching the designers did more "create" and "write" actions and when 3D modelling the designers did more "continue" actions which provided more detail in the co-created representation. The effect of facilitating more detailed digital representations is that the result of the remote 3D design sessions was a more developed design solution.

The second most notable result of this study was that the designers worked together continuously when sketching. They stayed focused on a common task. The designers in the CVE worked in two modes: together on the same task, and separately on individual tasks. The CVE provided awareness of the collaborator but also allowed the collaborating designers to have their own view of the design model and work independently, while checking in with each other occasionally.

## 3 Adding Tangible Interaction to 3D Design

The interaction with most CVEs is still enabled by the keyboard and mouse. Interaction technology is moving towards alternatives to the keyboard and mouse, such as pen interfaces, the Wii, direct brain-computer interaction, and tangible interaction. A study by Kim and Maher (2008) looks at the difference in the design process when designers use a mouse and keyboard vs 3D blocks as tangible input devices.

Tangible user interfaces (TUIs) offer physical objects as an alternative to typical computer input and output devices, and are often combined with augmented reality (AR) blending reality with virtuality (Azuma et al. 2001; Azuma 1997;

Dias et al. 2002). Using a tabletop system, designers can interact with 3D digital models more directly and naturally using TUIs while still utilizing verbal and non-verbal communication (Bekker 1995; Scott et al. 2003; Tang 1991). Many researchers have proposed that tangible interaction combined with AR display techniques might affect the way in which designers perceive and interact with digital models (Billinghurst and Kato 2002; Streitz et al. 1999; Ullmer and Ishii 1997). However, the central preoccupation of research on TUIs has been in developing new prototypes for design applications focusing on the functionality, where the Kim and Maher (2008) study provides empirical evidence for the effect of TUIs on designers' spatial cognition.

This empirical study considers the effects of TUIs on designers' spatial cognition using a protocol analysis. In order to focus on the impact of tangible interfaces on designers' spatial cognition, two settings are compared: a graphical user interface (GUI) as a desktop computer with a mouse and keyboard in ArchiCAD, and a TUI as tabletop system with 3D blocks in ARToolkit.

The tabletop system was developed at the University of Sydney and is described in (Daruwala, 2004). The tabletop system is shown in Figure 1 The horizontal display provides the means on which tangible interaction can take place and the binary patterns of the 3D blocks were made in ARToolKit[1] for the display of the 3D virtual models. A web camera captures the patterns and outputs them on a vertical LCD display in real time. In order to provide a same visual modality as the GUI environment, an LCD screen is used for the TUI session instead of a head mounted display (HMD).

Table 4 shows the set-up for the TUI session. The LCD screen was fixed to the left of the designers, and a 2D studio plan and 3D blocks were placed on the horizontal table. The web camera was set at a suitable height and angle to detect all the markers of the 3D blocks. A DVR (digital video recording) system was set to record two different views on one monitor, where one camera was used to monitor designers' behaviour and the other to capture the images on the LCD screen. This enabled the experimenter to observe designers' physical actions and the corresponding changes in the representation. A microphone was fed into the DVR system through a sound mixer and the camera filmed to record a clear view of designers' behaviours. A pair of designers sat at the same side of the table.

Table 5 shows the set-up for the GUI sessions. The overall experiment set-ups were similar to those of the TUI sessions. However, the camera was set to the left of the designers to avoid the LCD screen set to the front of the table from blocking its view of the designers. A pair of designers sat together in front of the computer, and the designer on the right usually operated the mouse.

Designers in the TUI sessions communicated design ideas by moving the objects visually, whereas designers in the GUI sessions discussed ideas verbally. Further, designers in the TUI sessions collaborated on handling the 3D blocks more interactively whereas designers in the GUI sessions shared a single mouse,

---

[1] ARToolKit is free AR software using a computer vision method and includes tracking libraries and source codes for the libraries, which is easy to use and allowed us to customise existing codes for our own applications (Billinghurst et al. 2003).

**Table 4.** Experimental set-up for the TUI session (Kim and Maher, 2008)

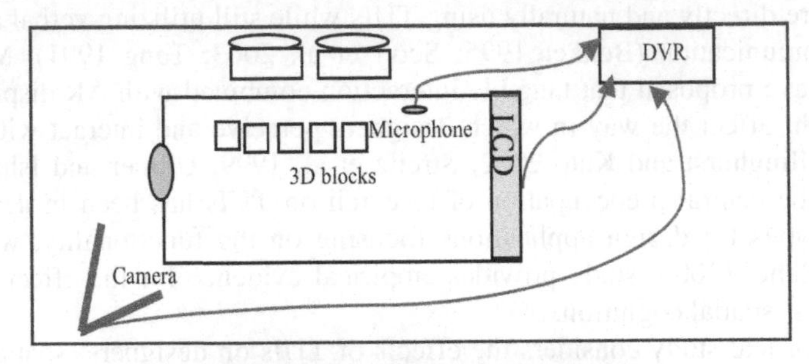

**Table 5.** Experiment setting for GUI session (Kim and Maher, 2008)

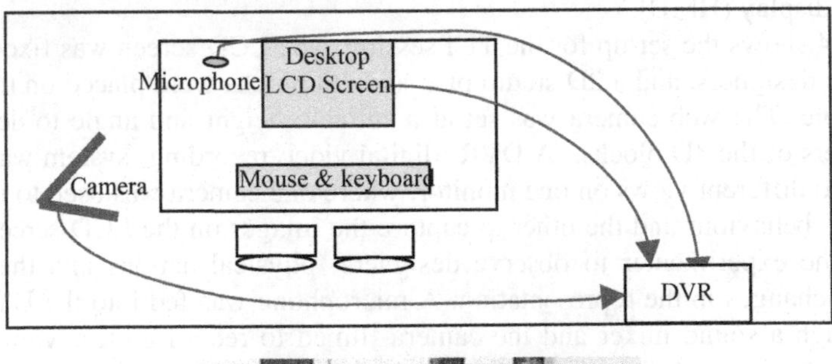

thus one designer mainly manipulated the mouse while the other explained what s/he was focusing on. These findings suggest that designers' collective interactions differed in the two design sessions.

It is notable that designers of the TUI sessions often kept touching the 3D blocks, and designers of the GUI sessions showed similar touching actions using the mouse. 'Touch' actions did not accompany any change in the placement of objects, but seemed to involve a cognitive process. Kim and Maher (2008) conjectured that 'Touch' gestures supported designers' perception of visuo-spatial features based on the argument by Kirsh and Maglio (1994): Some actions that appear unmotivated actually play valuable roles in improving performance, for instance, simplifying mental computation, from a perspective of epistemic goals.

Designers in the TUI sessions randomly placed pieces of furniture on the horizontal display of the plan, and then decided on their locations by moving them around. They were acting spontaneously, responding to their perceptual information straight away. On the other hand, designers in the GUI sessions seemed to decide their actions based on the information initially given rather than perceptual information. For example, regarding the placement of a new desk, designers in the GUI sessions emphasized the function of a desk for a computer programmer by saying "the programmer might need a desk for holding little computer things" and placing it in the corner. However, designers in a TUI session considered two locations for the desk, in the corner or near the window, then deciding to put it near the window so that the designer could look out, thus creating a spatial relationship between the desk and window. These findings suggest that designers developed design ideas in different ways according to the different interaction modes.

Through the results of the experiments, Kim and Maher (2008) found that the physical interaction with objects in TUIs produce epistemic actions as an 'exploratory' activity to assist in designers' spatial cognition. Further, the epistemic 3D modeling actions afforded by the interface off-load designers' cognition, and the naturalness of the direct hands-on style of interaction promote designers' immersion in designing, thus allowing them to perform spatial reasoning more effectively. In addition, designers' perception of visuo-spatial information, especially 'spatial relations', was improved while using the 3D blocks. The simultaneous generation of new conceptual thoughts and perceptual discoveries when attending to the external representation may also be explained by a reduction in the cognitive load of holding alternative design configurations in a mental representation.

In terms of the design process, designers' problem finding behaviours were increased in parallel with the change in designers' spatial cognition. The 'problem-finding' behaviours and the process of re-representation provided the designers with deeper 'insight' leading to key concepts for creative design. In summary, the study provides empirical evidence for the following views on TUIs: firstly, TUIs change designers' spatial cognition, and secondly, the changes of the spatial cognition are associated with problem finding behaviours typically associated with creative design processes.

## 4 Adaptive Virtual Worlds

While most developments and studies of CVEs focus on how to support collaboration, little research attention is given to the design of the virtual world as a place. Since a CVE is a virtual world, we can ascribe behaviours to the objects in the world in addition to designing their geometry and location in the world. This provides an opportunity to rethink how places are designed in CVEs. Gu and Maher

(2005) present an approach to designing places in CVEs that are responsive to the needs of the people in the world and automatically adapt to changing needs.

Typically, fixed behaviours are ascribed to 3D objects in a virtual world. Maher and Gero (2002) propose a multi-agent system, shown in Figure 1, to represent a 3D virtual world so that each object in the world has agency. With sensors and effectors as the interface to the 3D virtual world, each agent can sense the world, reason about the goals and modify the virtual world to satisfy the goals. 3D virtual worlds developed using this model can adapt their designs to suit different needs.

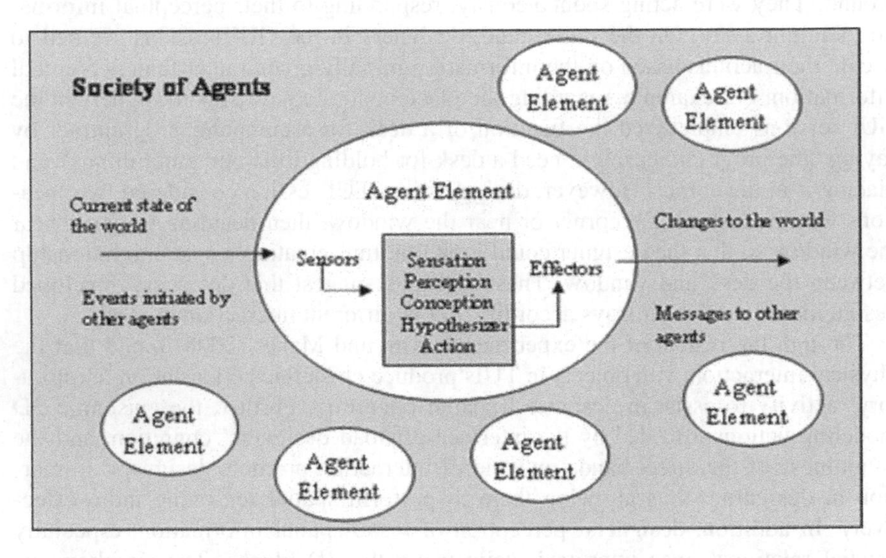


**Fig. 1.** A 3D virtual world as a multi-agent system (Maher and Gero 2002)

Maher and Gu (2005) developed this agent model further to automatically generate and adapt a 3D virtual world. The generative design agent (GDA) is an agent whose core processor is a design grammar. In addition to a world agent, each object in the world has agency and is capable of modifying itself. The rules in the grammar were grouped to provide the following design functions:

- Layout rules to identify functional spaces.
- Object rules to select and place the objects that enable the functions.
- Navigation rules to select and specify navigation methods.
- Interaction rules to select and ascribe behaviours to objects.

The GDA model enables 3D virtual worlds to be dynamically designed as needed. As the core of a GDA's design component, the generative design grammar includes the representation of design context of 3D virtual worlds in the forms of state labels, which can be used to match against the GDA's current interpretation for directing the grammar application. This research provides new insights for 3D virtual worlds from the following perspectives:

The GDA model introduces dynamics and autonomy to the designs of 3D virtual worlds. Virtual worlds designed with the GDA model do not have a static infrastructure like built environments. The CVE is designed for a particular "moment", and reflects its inhabitants' needs of that "moment".

The generative design grammar framework serves as a base for developing generative design grammars with different styles that suits different design purposes. This means that different agents can capture different styles that can be modified, combined, and evolved.

The generative design grammar framework also provides a foundation to formally study the styles of 3D virtual worlds. Compared to other novice designs, virtual worlds designed with a specific style in mind will achieve better consistency in terms of visualisation, navigation and interaction, and this consistency provides a strong base to assist its occupants' orientations and interactions in the virtual worlds.

## 5 Summary

This chapter provides an overview of the ways in which CVEs provide a unique approach to supporting collaborative design. The concept of a CVE differs from more traditional collaborative tools by providing an immersive sense of placc in a multi-user 3D virtual world. Following a presentation of the technical and social issues in using CVEs for collaborative design, the chapter provides an overview of three research projects that consider the impact of CVEs on the collaborative design process, the impact of new interaction technologies on the collaborative design process, and the potential for CVEs to provide a new kind of material for designing virtual places. The use of CVEs for collaborative design facilitates the seamless movement from working closely on the same task to working individually on a complex shared design task. The use of new interface technologies, such as tangible user interfaces, has the potential to redirect the focus of the designers on spatial relationships rather than the geometry of the parts, and to facilitate problem finding behaviour. Finally, as a new design material, CVEs allow designers to create proactive and adaptive places that respond to the changing needs of the avatars.

## References


Azuma, R., Baillot, Y., Behringer, R., Feiner, S., Julier, S., Maclntyre, B.: Recent advances in augmented reality. Computer Graphics and Applications 21(6), 34–47 (2001)

Azuma, R.T.: Survey on augmented reality. Presence: Teleoperators and Virtual Environments 6(4), 355–385 (1997)

Bekker, M., Olson, J.S., Olson, G.M.: Analysis of gestures in face-to-face design teams provides guidance for how to use groupware in design. In: The Symposium on Designing Interactive Systems DIS 1995, pp. 157–166 (1995)

Benford, S., Greenhalgh, C., Rodden, T.: Collaborative virtual environments. Communications of the ACM 44(7), 79–85 (2001)

Billinghurst, M., Kato, H., Poupyrev, I.: ARToolKit: a computer vision based augmented reality toolkit. In: IEEE VR 2000, New Jersey (2000)

Churchill, E.F., Snowdon, D.N., Munro, A.J. (eds.): Collaborative Virtual Environments: Digital Places and Spaces for Interaction. Springer, London (2001)

Clark, S., Maher, M.: Collaborative learning in a 3D virtual place: Investigating the role of place in a virtual learning environment. Advanced Technology for Learning 3(4), 208–896 (2006)

Daruwala, Y.: 3DT: Tangible input techniques used for 3D design & visualization, Masters Thesis, The University of Sydney (2004)

Dias, J.M.S., Santos, P., Bastos, L., Monteiro, L., Silvestre, R., Diniz, N.: MIXDesign: Tangible mixed reality for architectural design. In: 1st Ibero-American Symposium in Computer Graphics (SIACG) (2002)

Gabriel, G.C.: Computer Mediated Collaborative Design in Architecture: The Effects of Communication Channels on Collaborative Design Communication, PhD Thesis, Architectural and Design Science, Faculty of Architecture. University of Sydney, Sydney (2000)

Gero, J.S., McNeill, T.: An approach to the analysis of design protocols. Design Studies 19(1), 21–61 (1998)

Gu, N., Maher, M.L.: Dynamic designs of 3D virtual worlds using generative design agents. In: Martens, B., Brown, A. (eds.) Computer Aided Architectural Design Futures 2005, pp. 239–248. Springer, Dordrecht (2005)

Gul, L.F., Maher, M.L.: Co-creating external design representations: comparing face-to-face sketching to designing in virtual environments. CoDesign International Journal of CoCreation in Design and the Arts 5(2), 117–138 (2009)

Kim, M.J., Maher, M.L.: The impact of tangible user interfaces on spatial cognition during collaborative design. Design Studies 29(3), 222–253 (2008)

Kirsh, D., Maglio, P.: On distinguishing epistemic from pragmatic action. Cognitive Science 18, 513–549 (1994)

Kvan, T.: Collaborative design: What is it? Automation in Construction 9(4), 409–415 (2000)

Kvan, T., Candy, L.: Designing collaborative environments for strategic knowledge in design. Knowledge-Based Systems 13, 429–438 (2000)

Maher, M.L., Rosenman, M., Merrick, K., Macindoe, O.: DesignWorld: An augmented 3D virtual world for multidisciplinary collaborative design. In: Kaga, A., Naka, R. (eds.) CAADRIA 2006: Proceedings of the 11th Conference on Computer-Aided Architectural Design Research in Asia, Osaka, Japan, pp. 133–142 (2006)

Maher, M.L., Bilda, Z., Marchant, D.: Comparing collaborative design behavior in remote sketching and 3D virtual worlds. In: Proceedings of International Workshop on Human Behaviour in Designing, Key Centre of Design Computing and Cognition, pp. 3–26. University of Sydney, Sydney (2005)

Maher, M.L., Gero, J.S.: Agent models of 3D virtual worlds. In: ACADIA 2002: Thresholds, California State Polytechnic University, Pomona, pp. 127–138 (2002)

McGrath, A., Prinz, W.: All that is solid melts into software. In: Churchill, E.F., Snowdon, D.N., Munro, A.J. (eds.) Collaborative Virtual Environments: Digital Places and Spaces for Interaction. Springer, London (2001)

Redfern, S., Naughton, N.: Collaborative virtual environments to support communication and community in internet-based distance education. Journal of Information Technology Education 1(3), 201–211 (2002)

Saad, M., Maher, M.L.: Shared understanding in computer-supported collaborative design. CAD Journal B(3), 183–192 (1996)

Scott, S.D., Grant, K.D., Mandryk, R.L.: System guidelines for co-located, collaborative work on a tabletop display. In: Proceedings of the European Conference Computer-Supported Cooperative Work, pp. 159–178 (2003)

Streitz, N.A., Geißer, J., Holmer, T., Konomi, S., Muller-Tomfelde, C., Reischl, W., Rexroth, P., Seitz, P., Steinmetz, R.: i-Land: An interactive landscape for creativity and innovation. In: The Conference on Human Factors. Computing Systems (CHI 1999), pp. 120–127 (1999)

Tang, J.C.: Findings from observational studies of collaborative work. International Journal of Man-Machine Studies 34, 143–160 (1991)

Ullmer, B., Ishii, H.: The metaDESK: Models and prototypes for tangible user interfaces. In: Proceedings of User Interface Software and Technology (UIST) (1997)

# Visual-Spatial Learning and Training in Collaborative Design in Virtual Environments

Maria Kozhevnikov[1] and Andre Garcia[2]

[1] Harvard Medical School and National University of Singapore,
[2] George Mason University, USA

**Abstract.** This chapter reviews different types of immersive virtual environments (IVE) and discusses the major advantages that these environments can offer in the domain of visual-spatial learning, assessment, and training. Overall, our review indicates that immersion might be one of the most important aspects to be considered in the design of learning and training environments for visual-spatial cognition. Furthermore, we suggest that only immersive virtual environments can provide a unique tool for assessing and training visual-spatial performance that require either the reliance on non-visual cues (motor, vestibular, or proprioceptive) or the use of egocentric frames of references.

**Keywords:** collaborative virtual environments, visual-spatial cognition, immersion, spatial transformations, spatial navigation.

## 1 Overview of Different Types of Virtual Environments

Virtual Environments (VEs) can be generalized as a class of computer simulations pertaining to a representation of three dimensional (3D) space and a human computer-interaction in that space (Cockayne & Darken, 2004). In order to produce an immersive virtual environment (IVE), in which the user perceives being surrounded by a 3D world, Virtual Reality (VR) technology combines real-time computer graphics (CG), body tracking technology, audio/video, touch and other sensory input and output devices. In this chapter, we will briefly discuss different types of IVEs used in the field of visual-spatial cognition research, followed by discussion of the major advantages that IVEs can offer for the domains of visual-spatial learning, assessment and training.

There are several common implementation of IVEs. The first of these is the head-mounted display (HMD). Most HMDs include a visual display worn on the head and over the eyes they covers the user's entire visual field, and a head tracking device that continuously updates the user's head position and orientation in the physical environment. A second type of commonly used virtual environment is the Cave Automated Virtual Environments (CAVE), where images are projected on the walls, floor, and ceiling of a room that surrounds a viewer. A third type of

X. Wang & J.J.-H. Tsai (Eds.): Collaborative Design in Virtual Environments, ISCA 48, pp. 17–26.
springerlink.com 

commonly used IVE is an Augmented or Mixed Reality environment that permits users to see and interact with virtual objects generated within the real physical environment. It is usually achieved by using live video feed that is digitally processed and "augmented" by the addition of computer-generated graphics (Blade & Padgett, 2002).

There are other types of VEs used in visual-spatial cognition research which combine the use of immersive technology and physical systems such as flight simulators, driving simulators, or omni-directional treadmills (walking simulators). These provide coherent multi-sensory environments for the user to perceive and explore, either through perceived vehicle or own-body movement. Recently, rather than using stationary platforms, driving simulators have incorporated rotating platforms that allow for different dimensions of motion (yaw, pitch, or roll motion). There has also been a distinction made between simulators that allow active versus passive transport (Cockayne & Darken, 2004). While active transport captures deliberate physical movements executed by the user (e.g., walking or running) and directly translates these into virtual movement, passive transport is not directly executed by the user, but by a vehicle which the user operates, and is carried by (e.g., a plane or wheelchair). This distinction is crucial due to the specific demands placed on the body during active locomotion versus passive locomotion such as limb movement, balance, coordination, and several other factors. An example of active transport is the omni-directional treadmill (ODT), which allows the participant to walk in any direction indefinitely, while all his/her movements are tracked. Driving and flight simulators are examples of passive transport simulators.

The above-mentioned VEs are currently among the most commonly used technologies for research, training, and assessment. Sometimes they are used in conjunction with one another and other devices. Examples of different IVEs can be found in the reviews of Darken & Peterson (2002), Loomis et al. (1999), and Wiederhold & Rizzo (2005).

## 2 Applications of VEs for Individual and Collaborative Activities

There are many benefits of IVEs over conventional 2D or non-immersive, 3D desktop VR environments for individual and collaborative activities. On the most general level, IVEs can be created to present simulations that assess and train human processes and performance under a range of stimulus conditions that are not easily (or safely) deliverable in non-virtual world conditions (Kidd & Monk, 2008; Rizzo & Schultheis, 2002). Furthermore, IVEs can be used for team performance assessment and training, particularly when tasks occur infrequently (competitive sporting events), are expensive to replicate (flight simulation) or present dangerous scenarios in hostile (military) environments (Salas et al., 2002). Since 1929, flight simulators have been used for safely acquiring expertise in a dangerous task, allowing trainers to learn the differences between experts and novices, and then help improve the performance of novices to get to expert levels more efficiently.

Experimental studies confirmed that flight simulators are effective in distinguishing expert from novice pilots by their flight performance and situation awareness (e.g., Jodlowski et al., 2003). Overall, VEs are extensively used in air and space applications, such as training, prototyping, assembly planning. NASA, for example, successfully employed VE technology in preparing astronauts for the Hubble Space Telescope repair mission, or in training air traffic controllers (Homan, 1994).

Interactive Virtual Environments are also applied to medical training. Verner et al (2003) used 3D medical simulation devices to identify the differences between expert surgeons and medical students, finding that experts exhibit more efficient hand movements throughout task performance. When using simulation and VR for training, Colt et al. (2001) found that novices who were trained for eight hours nearly caught up to the performance level of expert surgeons in performing a flexible fiberoptic bronchosocopy.

In general, IVEs offer the potentials to create systematic human testing, training, and treatment environments that allow for the precise control of complex, immersive, and dynamic 3D visual presentations (Wiederhold & Rizzo, 2005). At the same time they allow sophisticated interaction, behavioural tracking, and performance recording. As we will discuss below, there are two major components of IVEs that might be particularly relevant to visual-spatial cognition. The first one is immersion, which allows for *egocentric* navigation (the user is surrounded by the environment), rather than *exocentric* navigation (the user is outside the environment, looking in). The second important factor is motor, vestibular (i.e., information from the inner ear about head and gross bodily movement in 3D space) and proprioceptive (i.e., information from muscles and joints about limb position) feedback. Whether the environments are driving simulators, flight simulators, HMDs, ODTs, or any type of VR technology, they can provide the user with different degrees of motor, vestibular, and proprioceptive feedback, which distinguishes them from non-immersive environments.

## 3 Design Applications for Visual-Spatial Cognition

Visual-spatial cognition research has already extensively incorporated VR technology (e.g., Chance et al., 1998; Darken & Sibert, 1996; Klatzky et al., 1998; Kozhevnikov et al., 2008, Kozhevnikov, 2009; Richardson, Montello, & Hegarty, 1999) due to the ease with which one can create a complex environment for participants to explore as well as record their behavior (see Loomis et al., 1999; Peruch & Gaunet, 1998 for a review).

However, with regards to assessment, learning and training, although these 3D virtual environments are both more appealing to the user and richer in object and spatial information, research thus far has not reached an entirely uniform conclusion regarding their effectiveness in promoting visual-spatial learning and task performance (e.g., Haskell & Wickens 1993; Van Orden & Broyles 2000). For instance, Van Orden and Broyles (2000) investigated how different types of displays affect performance on a variety of flight control tasks and found that 3D displays only increased performance on a collision avoidance task. The researchers suggested that 3D representations of data might be beneficial primarily for tasks requiring the integration and prediction of motion within limited spatial areas. Thus, the question still remains of how and when these 3D environments are beneficial for visual-spatial learning, assessment, and training. To answer this question, we will review the research on the utility of IVEs in two fields of visual-spatial cognition research: 1) locomotion and spatial navigation and 2) 3D spatial transformations.

### 3.1 Locomotion and Spatial Navigation

As it is now recognized in visual-spatial cognition literature, optic flow (information from the visual system signaling how fast the visual world is moving past the eyes) and 3D perceptual representations together do not explain sufficiently the control of locomotion and spatial navigation performance (e.g., Loomis & Beall, 1998). Visually controlled locomotion is often accomplished with supplementary non-visual information about the person's motion, such as vestibular and somatosensory signals, which provide the operator of a vehicle with information about vehicle velocity and acceleration (Gillingham & Wolfe, 1986) and, in case of flying at night, provide information about aircraft orientation (Loomis & Beall, 1998).

Furthermore, recent psychophysical studies revealed an unexpectedly important contribution of vestibular cues in distance perception and steering, prompting a re-evaluation of the role of vestibular interaction in driving simulation studies (Kemeny & Panerai, 2003). For instance, Macuga et al (2007) showed that inertial information from real vehicle motion is critical for eliminating errors during lane changing found in other experiments performed in driving simulators with fixed motion base. Overall, the results of the driving simulator studies (Kemeny & Panerai, 2003) recommend the use of a large field of view (enhancing to the feeling of immersion) and a moving platform (allowing vestibular and inertial feedback) while considering the design of driving simulators.

Taking into account the importance of vestibular and somatosensory proprioceptive information in visually controlled locomotion, it is not surprising that different VEs that allow active or passive body transport such as HMDs or driving simulators with rotating platforms, can provide unique environments for research, assessment, and training in relation to locomotion activities.

Spatial navigation is another cognitive activity in which training and assessment via IVEs might prove worthwhile. Recent studies demonstrated that spatial navigation relies on two distinct processes, piloting and path integration (Loomis et al., 1998). Piloting relies on position fixing based on environmental cues such as landmarks, while path integration refers to the ability to integrate self-motion information to estimate one's current position and orientation relative to the origin. The difference between path integration and piloting is that piloting requires an internal representation of a route that has been traversed; thus containing information about various path segments and turns as well as off-route landmarks. In contrast, the representation underlying path integration is a continually updated abstraction derived from computations on route vector information (Loomis et al., 1998).

Many successful navigators seem to rely on path integration strategies (Blajenkova et al., 2005; Kozhevnikov et al, 2006) rather than piloting, which is the only navigational strategy available in unfamiliar environments, or under conditions where the visual information is scarce. Path integration, unlike piloting, depends not only on visual, but also on motor, vestibular and proprioceptive cues (to estimates one's current position (Loomis et al., 1998; Bigel & Ellard, 2000). Therefore, taking into account the importance of such non-visual cues, one can conclude that IVEs might provide much more effective research and assessment tool to study navigational strategies in comparison with 2D conventional non-immersive displays. Specifically, IVEs might be hepful in isolating path integration and piloting navigational strategies. For example, IVEs make it possible to study path integration based on visual input alone, by providing sufficient optic flow information for sensing self-motion while eliminating any environmental position cues (e.g., Klatzky et al., 1998). In contrast, one can use an IVE to investigate piloting without any path integration (see Loomis et al., 1999 for a review). An important advantage of IVEs over real environments in path integration research is that the experimenter can be assured that self-localization is not mediated by any incidental positional cues, such as landmarks, and only those cues that are provided in the 3D spatial database are available to the participant.

### 3.2 3D Spatial Transformations

Recent psychological and neuroscience research on visual-spatial cognition suggests dissociation between processes that require the mental manipulation of objects from a stationary point of view (*allocentric* spatial transformations) and processes that require the imagining taking different perspectives in space (*egocentric* spatial transformations) (e.g., Easton & Sholl, 1995; Rieser, 1989). Allocentric manipulation of objects or an array of objects (e.g., mental rotation of cubes or other geometrical figures) involves imagining movement relative to an object-based frame of reference, which specifies the location of one object (or its parts) with respect to other objects. In contrast, egocentric transformation such as imagining a different orientation (perspective) involves movement of the egocentric frame of reference, which encodes object locations with respect to the front/back, left/right and up/down axes of the observer's body. The encoding of visual-spatial stimuli in relation to egocentric (body-centred) spatial frames of reference has been shown to be critical for successful performance in many real-world tasks, such as real-world navigation on the land, air and water, scene encoding, remote control, weapon deployment, etc. (Aoki et al., 2008; Kozhevnikov et al., 2007).

Several research studies revealed that immersion is necessary to provide adequate information for building the spatial reference frame required for egocentric encoding and transformations (Kozhevnikov 2008; Kozhevnikov, 2009), and that only IVEs can provide reliable environments for measuring and training aspects of visual-spatial ability related to egocentric spatial transformations. For instance, Kozhevnikov et al. (2007) designed a 3D immersive perspective-taking ability (3DI PTA) test as a measure of egocentric spatial transformation ability and compared subjects' performance on this test in IVE, traditional 2D, and non-immersive 3D (stereoscopic glasses) environments. In the IVE condition, the test was presented via HMD (see Figure 1), while in traditional 2D and 3D non-immersive environments, spatial scenes were presented to the participant on a standard computer monitor. On each trial, the participant was placed in a location inside the scene in IVE or shown the scene exocentrically in 2D and 3D non-immersive environments. The participants were explicitly instructed to imagine taking the perspective of the avatar located in the array of objects and then to point to a specific target from the imaginary perspective of the avatar by using the pointing device or joystick.

Comparative analysis of subjects' responses in the three environments revealed that the IVE (3D PTA) is a unique instrument to measure egocentric spatial ability. Specifically, while the participants were as accurate on 3D PTA as on the 2D PTA and 3D desktop PTA, their errors were qualitatively different. In particular, in 3D PTA, most errors were systematically due to confusion between "right-left" and "back-front" coding in respect to the body indicating that they indeed were relying more on body-centred frame of reference. In contrast, in 2D and 3D desktop non-immersive environments, participants made more "allocentric" errors characterized by over-rotating or under-rotating the scene. Furthermore, it has been shown that the 3-D PTA test had a significantly stronger training effect than the tests administered in the two other environments.

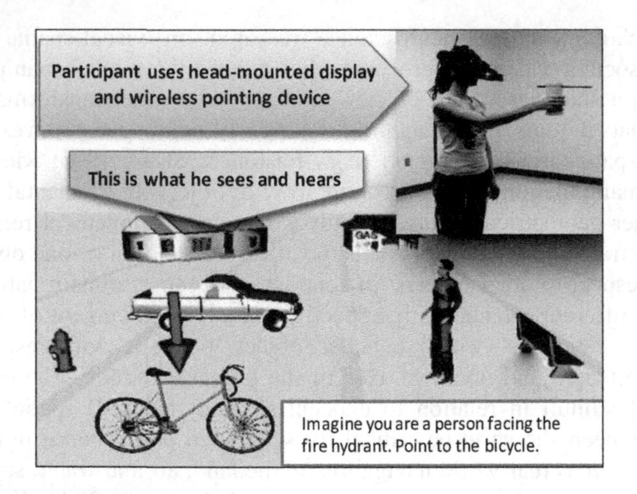


**Fig. 1.** Perspective-taking ability test administered in IVE.

A similar pattern of the use of egocentric frame of reference in IVE was found in another study (Kozhevnikov et al., 2008) for mental rotation tasks. The study compared subjects' performance on mental rotation tasks within traditional IVE, 2D, and 3D desktop non-immersive environments. In this study, participants were asked to mentally rotate 3D objects along the picture (X), vertical (Y), or depth (Z) axes (Figure 2).

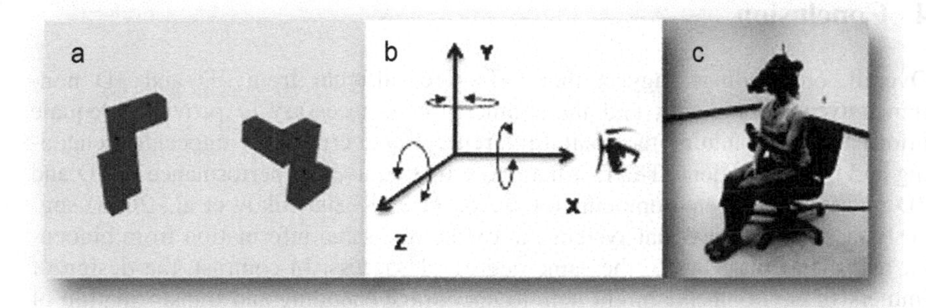


**Fig. 2.** Mental rotation test: a) Example item, which includes two 3D shapes that have to be mentally rotated into alignment, b) Three principle axes of rotation, c) Test in 3DI environment, which includes HMD with position tracking.

While the patterns of subjects' responses were not significantly different in 2D and 3D desktop environments, we found a unique pattern of responses in the IVE environment, suggesting that immersion triggered significantly greater use of an egocentric frame of reference (specifically retinocentric frame) than the two other non-immersive environments. In particular, in IVE, the rate of rotation in depth (around Z axis) was significantly slower than the rate of rotation in the picture plane (around the X or Y axes). This suggests that the subjects were in fact rotating 2D retina-based representations, since the rotation in depth is more difficult than in the picture plane due to foreshortening and occlusion. However, in 2D and 3D non-immersive environments, the rates of mental rotation around the X and Z axes were identical. This suggests that non-immersive displays encourage the use of more "artificial" encoding and transformation strategies, where the objects' components are encoded in terms of "vertical" and "horizontal" relations with regard to their own internal structure, as well as to the sides of the computer screen.

The use of different spatial frames of reference in immersive and immersive environments (and reliance on egocentric spatial frame of references in immersive environments) would also explain why the results of the training studies show no transfer from training in 2D environments to performance in IVE or to the real world. For instance, Pausch et al. (1997) reported that IVE users who practiced first with conventional 2D displays in visual search tasks did not show any improved performance in IVE, but not vice versa. This implies that using desktop graphics to train users for real world search tasks might not be efficient, and may actually be counterproductive. The reason for this effect, we suggest, is that the encoding of spatial relations and the cognitive strategies applied to perform visual-spatial transformations in 2D non-immersive and IVEs are different. We suggest that IVEs with a variety of simulated 3D stimuli will provide the most efficient environment for training visual-spatial skills that will generalize and transfer to real-world tasks.

## 4 Conclusion

Overall, our findings suggest that IVEs are different from 2D and 3D non-immersive environments, and that immersion is necessary to provide adequate information for building the spatial reference frame crucial for egocentric encoding and transformations. The fact that there was equivalent performance in 2D and 3D non-immersive environments on spatial task (Kozhevnikov et al., 2008) suggests that the human visual system can extract the same information from binocular and monocular cues to the same degree of success. In contrast, the design of immersive environments might help to encourage encoding and transformation of an image with respect to the egocentric spatial frame of reference, similar to a real environment, as well as provide non-visual cues that are important for locomotion and navigation.

Thus, only an IVE can provide a unique and more realistic environment for assessing and training visual-spatial skills that require either non-visual cues (e.g., vestibular or proprioceptive) or the use of egocentric frames of reference. Therefore, immersion might be one of the most important aspects to be considered in the design of learning and training environments for visual-spatial cognition.

## Acknowledgements

We would like to acknowledge Michael Becker and Olesya Blazhenkova for helpful comments and their time and contributions and Office of Naval Research (grant ONR_ N000140611072) for support.

## References

Aoki, H., Oman, C., Buckland, D., Natapoff, A.: Desktop-VR system for preflight 3D navigation training. Acta Astronautica 63, 841–847 (2008)

Bigel, M., Ellard, C.: The contribution of nonvisual information to simple place navigation and distance estimation: An examination of path integration. Canadian Journal of Experimental Psychology/Revue canadienne de psychologie expérimentale 54(3), 172–185 (2000)

Blade, R.A., Padgett, M.L.: Virtual environments standards and terminology. In: Stanney, K. (ed.) Handbook of Virtual Environments: Design, Implementation, and Applications, pp. 15–27. Lawrence Erlbaum Associates Publishers, Mahwah (2002)

Blajenkova, O., Motes, M., Kozhevnikov, M.: Individual differences in the representations of novel environments. Journal of Environmental Psychology 25(1), 97–109 (2005)

Chance, S., Gaunet, F., Beall, A., Loomis, J.: Locomotion mode affects the updating of objects encountered during travel: The contribution of vestibular and proprioceptive inputs to path integration. Presence 7(2), 168–178 (1998)

Cockayne, W., Darken, R.: The application of human ability requirements to virtual environment interface design and evaluation. In: The handbook of task analysis for human-computer interaction, pp. 401–421 (2004)

Colt, H.G., Crawford, S.W., Gaulbraith, O.: Virtual reality bronchoscopy simulation. Chest 120, 1333–1339 (2001)

Darken, R., Peterson, B.: Spatial orientation, wayfinding, and representation. In: Stanney, K. (ed.) Handbook of Virtual Environments: Design, Implementation, and Applications, pp. 493–518. Lawrence Erlbaum Associates Publishers, Mahwah (2002)

Darken, R., Sibert, J.: Navigating large virtual spaces. International Journal of Human-Computer Interaction 8(1), 49–71 (1996)

Easton, R., Sholl, M.: Object-array structure, frames of reference, and retrieval of spatial knowledge. Journal of Experimental Psychology: Learning, Memory, and Cognition 21(2), 483–500 (1995)

Gillingham, K.K., Wolfe, J.W.: Spatial Orientation in Flight (Technical Report USAF-SAM-TR-85-31), Brooks Air Force Base, TX: USAF School of Aerospace Medicine, Aerospace Medical Division (1986)

Haskell, I., Wickens, C.: Two- and three-dimensional displays for aviation: A theoretical and empirical comparison. International Journal of Aviation Psychology 3(2), 87–109 (1993)

Homan, W.J.: Virtual reality: Real promises and false expectations. EMI: Educational Media International 31(4), 224–227 (1994)

Jodlowski, M.T., Doane, S.M., Brou, R.J.: Adaptive expertise during simulated flight. In: Proceedings of the Human Factors and Ergonomics Society. In: 47th Annual Meeting, HFES, Denver, Colorado (October 2003)

Kemeny, A., Panerai, F.: Evaluating perception in driving simulation experiments. Trends in Cognitive Sciences 7(1), 31–37 (2003)

Kidd, D.G., Monk, C.A.: The effects of dual-task inference and response strategy on stop or go decisions to yellow light changes. In: Proceedings of the 5th International Symposium on Human Factors in Driving Assessment, Training, and Vehicle Design, Big Sky, Montana (June 2009)

Klatzky, R., Loomis, J., Beall, A., Chance, S., Golledge, R.: Spatial updating of self-position and orientation during real, imagined, and virtual locomotion. Psychological Science 9(4), 293–298 (1998)

Kozhevnikov, M.: The role of immersive 3-D environments in mental rotation. Paper was be presented at the Psychonomic Society 50th Annaul Meetihg, November 12-22 (2009)

Kozhevnikov, M., Blazhenkova, O., Royan, J., Gorbunov, A.: The role of immersivity in three-dimensional mental rotation. In: Paper was presented at third International Conference on Design Computing and Cognition DCC 2008, Atlanta, GA (2008)

Kozhevnikov, M: Individual difference in allocentric and agocentric spatial ability, Technical report, Office on Naval Research, N000140611072 (2007)

Kozhevnikov, M., Motes, M., Rasch, B., Blajenkova, O.: Perspective-taking vs. mental rotation transformations and how they predict spatial navigation performance. Applied Cognitive Psychology 20, 397–417 (2006)

Loomis, J., Beall, A.: Visually controlled locomotion: Its dependence on optic flow, three-dimensional space perception, and cognition. Ecological Psychology 10(3), 271–285 (1998)

Loomis, J., Blascovich, J., Beall, A.: Immersive virtual environment technology as a basic research tool in psychology. Behavior Research Methods, Instruments & Computers 31(4), 557–564 (1999)

Loomis, J., Klatzky, R., Golledge, R., Philbeck, J.: Human Navigation by Path Integration, Wayfinding Behavior: Cognitive Mapping and Other Spatial Processes. Johns Hopkins University Press, Baltimore (1998)

Macuga, K.L., Beall, A.C., Kelly, J.W., Smith, R.S., Loomis, J.M.: Changing lanes: inertial cues and explicit path information facilitate steering performance when visual feedback is removed. Experimental Brain Research 178, 141–150 (2007)

Pausch, R., Proffitt, D., Williams, G.: Quantifying immersion in virtual reality. In: SIGGRAPH (August 1997)

Péruch, P., Gaunet, F.: Virtual environments as a promising tool for investigating human spatial cognition. Cahiers de Psychologie Cognitive/Current Psychology of Cognition 17(4), 881–899 (1998)

Richardson, A., Montello, D., Hegarty, M.: Spatial knowledge acquisition from maps and from navigation in real and virtual environments. Memory & Cognition 27(4), 741–750 (1999)

Rieser, J.: Access to knowledge of spatial structure at novel points of observation. Journal of Experimental Psychology: Learning, Memory, and Cognition 15(6), 1157–1165 (1989)

Rizzo, A., Schultheis, M.: Expanding the boundaries of psychology: The application of virtual reality. Psychological Inquiry 13(2), 134–140 (2002)

Salas, E., Oser, R., Cannon-Bowers, J., Daskarolis-Kring, E.: Team training in virtual environments: An event-based approach. In: Stanney, K. (ed.) Handbook of Virtual Environments: Design, Implementation, and Applications, pp. 873–892. Lawrence Erlbaum Associates Publishers, Mahwah (2002)

Van Orden, K., Broyles, J.: Visuospatial task performance as a function of two- and three-dimensional display presentation techniques. Displays 21(1), 17–24 (2000)

Verner, L., Oleynikov, D., Holtmann, S., Haider, H., Zhukov, L.: Measurements of the level of surgical expertise using flight path analysis from da Vinci Robotic Surgical System. Medicine Meets Virtual Reality 11, 373–378 (2003)

Wiederhold, B., Rizzo, A.: Virtual reality and applied psychophysiology. Applied Psychophysiology and Biofeedback 30(3), 183–185 (2005)

# Part II
# Representation and Embodiments in Collaborative Virtual Environments: Objects, Users, and Presence

Design Representation and Perception in Virtual Environments
Chiu-Shui Chan (Iowa State University)

Design Paradigms for the Enhancement of Presence in Virtual Environments
Rivka Oxman (Technion Israel Institute of Technology)

Co-presence in Mixed Reality-Mediated Collaborative Design Space
Xiangyu Wang and Rui Wang (University of New South Wales)

# Design Representation and Perception in Virtual Environments

Chiu-Shui Chan

Iowa State University, USA

**Abstract.** Two important cognitive activities involved in designing in virtual environments are explored in this chapter. The first activity is design representation that is mentally created during the design processes. In virtual environments, particularly the full-scale immersive virtual reality settings, the nature of representation applied to design generation is different from the one applied in the conventional design environments such as pencil-and-paper or physical model-making mode. The second activity relates to human perception, which has not been changed by high-tech developments. Perception in virtual environments provides information to allow designers to understand the environmental impact generated from design. Additional knowledge of media applications and their corresponding representations has created new definitions of identity and privacy, which also has created interesting design impacts, subtle cultural effects, and social interactions. These phenomena are described through examples in this chapter.

**Keywords:** virtual reality, mental representation, perception, design cognition.

## 1 Background

"Design representation" has been explored and discussed in a number of studies (Simon, 1969; Chan et al., 1999; Eastman, 2001; Goldschmidt, 2004), covering: (1) how representation is created, (2) how it is used for design in the conventional design environments, and (3) the uses of traditional design modes in creating artefacts; It provides the field of design with a comprehensive understanding of its relationships with design thinking.

Along with the emerging new media and the rapid changes of information technology, digital design has become a leading trend, and design thinking in the digital world has changed accordingly. Representation applied in the digital design world has been modified to meet the different design situations. The newly emerged virtual environments, with the advantages of visualizing the virtual world through perception, have created different dimensions of representation for design. Thus, the representation and perception are two important human cognitive faculties involved with design in virtual environments.

This chapter develops these ideas and discusses: (1) the changed nature of representation utilized in virtual environments, (2) the type of information that virtual

X. Wang & J.J.-H. Tsai (Eds.): Collaborative Design in Virtual Environments, ISCA 48, pp. 29–40.
springerlink.com 

environments provide through perception, and (3) how the representation and perception of design in virtual environments will impact societal cognition to shape a new sub-culture. *Perception* is the ability to convert the presented sensory stimuli in the world into organized psychological apprehension, or the human function of interpreting sensory information to get immediate/direct experiences of the world (Chan, 2008).

Virtual environments are defined as the digital worlds displayed through projectors, which did not exist before the 19th Century. It is necessary to differentiate between "representation" and "media". Strictly speaking, if there is something used to display a design idea or to transform a concept from the internal mind to the external world, it is a design media. Media, from the field of mass communication and-advertisement, is simply a means of communication. However, when the thing is used for creating a design, it is a design representation.

It should be noted that some images used in this chapter are merely still pictures taken from studies done in the virtual worlds, which are not capable of displaying the dynamic and run-time characteristics of representation and perception in virtual environments; however, they do illustrate the basic concepts.

## 2 Representation and Media

*Representation* is an entity used to represent something else (Hesse, 1966). The meaning of representation varies across research contexts. In the area of artificial intelligence, representation is used inclusively with the representation of a problem (problem representation), which includes the initial encoding of the problem, data structures for storing related information, rules for drawing references about the problem, and heuristics that lead the search to a solution (Fink, 2003). In design, designers use suitable means to mentally create design concepts, apply communication channels (media) to express their design concepts and turn the concepts into external visible artefacts (products); so that designers and other viewers (or clients) can visualize the design in progress. These various means used for creation are internal representations, whereas the artefacts are external representations of the design. There are five categories of design communication media that have been commonly used.

1. The oldest and historical one is the *pencil-and-paper* mode. Results generated from this mode are usually abstract drawings, quick sketches, or even constructional (or working) drawings.
2. During the Renaissance Period (1400), handmade *physical models* were introduced to generate 3D objects for study or display.
3. When the digital computer was created in 1937 and *graphic/modelling software* was developed, designers used appropriate software to generate digital drawings or models.
4. When cameras became available, *film and video* were also applied to create animation for showing/generating design concepts, or for demonstrating design products/processes.
5. Lastly, *Virtual Reality* (VR) facilities emerged as a class of advanced media for visualization, simulation, and recreational purposes.

Media are tools which require unique mental operations, procedures, techniques, and representations to convert concepts into forms. These unique mental operations are parts of *design cognition,* defined as the human abilities or intelligences to organize design information and problem structure for creating man-made artefacts. The human mind is regarded as the unit of information processing. When using different media, different representations and cognitive processes are utilized, which can be described as layers of mental organizing information.

## 3 Layers of Organizing Information in the Design Processes

Described by psychological studies done in the field of architectural design (Akin, 1986; Rowe, 1987; Chan, 1990), the design processes have the following characteristics: designers would stylistically choose (Chan, 2001a) the required design information given by the clients or information gathered through research on architectural functionalities suitable for the project. These pieces of information ultimately become design constraints for developing design strategically and for creating tectonic elements, which signify the first layer of *design information.*

Design strategies are then developed, based on the selected design requirements and functionalities to be addressed. Accordingly, designers will also create their own personal design constraints methodologically to meet the functional requirements and to narrow down the problem space. Sequentially, designers will draw from memory or from the layer of design information to generate a solution. Thus, certain design ideas, which might be images, diagrams, or abstract concepts, are created in designers' mind's eye, which are the *internal representations.* These internal representations are results of cognitive operations that turn the design information into concepts. Data that could symbolize the cognitive processes for creating the internal representations constitutes the *cognitive information.*

After design concepts are created and internal representations are developed, designers apply certain procedures through selected media to make them visible. Each media has its own operational world with unique methods of operations. Different media require different algorithms or procedures for operation. These algorithms or techniques define another layer of *media information*: (1) if the media is pencil-and-paper mode, drawing skills in the use of brushes, colours, proportions, and compositions are essential; (2) if the media is computer software, the comments and functions of that software for creating 3D forms or digital shapes are necessary; (3) using film or video for design requires a different set of methodologies to perform the task, because the syntax composition and semantic language in film production are unique; (4) design in Virtual Reality have requirements on the format of virtual models in order to make them accurately displayed in 3D stereoscopic mode.

The entire design process phenomena consist of putting information through representation to create design concepts, and utilizing media to make the concepts visible until the final solution is reached. These complicated mental processes will generate some *external representation* of a drawing, video, physical model, digital model, virtual model; or the combination of all.

After the design is completed and constructed, the built (or to-be-built) environment yields information in space affecting human cognitive performance positively or negatively. Understanding what the occupants' actions and reactions are, as well as how and why they occur in the environment will benefit the designers' design and the occupants' cognitive performance. This layer of post-occupancy information can be visualized through virtual environments and obtained by perception.

## 4 Impacts of Media to Representation

Design processes are reflective ones (Schon, 1983). Designers are always adjusting their media information to map the internal representation with the external one for satisfying some functional and goal requirements until the results are accomplished. Among these processes, the internal representation is flexible, for it has an abstract nature and is an isomorphism of the external representation. Designers also cognitively generate a problem representation to solve the design problem at hand. Designers' responsibilities and cognitive activities are constantly converting the abstract internal representation into concrete external ones by continuously transforming mental data into external form, and perceiving results to determine sequential moves. Studies found even when an engineering task is very simple, experts and novices construct problem representation differently (Larkin, 1985).

Similarly, different designers use different problem representations across different media. Representations created in virtual environments are different from others due to their complicated media nature. For example, if the design conducted in virtual environments is a real world problem dealing with generating artifacts to satisfy certain issues in reality, then the representations used are mental images mirroring the realistic images reflected from perception, which means that these mental representations are visually driven. On the other hand, if the design is just a creation of artifacts, and not a real world problem, then the representations created are arbitrary images coming from imagination. Two experiments conducted in two immersive virtual reality settings clearly explained these differences, and immersive virtual reality has the following attributes that could justify the validity of these two hypotheses.

## 5 The Nature of Immersive Environments

VR is a technology that simulates objects and spaces through 3D computer-generated models. In a VR model, the feeling of realism is derived from a sequence of high-resolution, stereoscopic images. If the display allows viewers to project themselves into the scene, then a virtual environment is created. If the scene is shown in full scale and viewers are surrounded by 3D images, an immersive environment is generated. In the immersive environments of the C2/C4 (both are three-sided) and C6 (a six-sided) at Iowa State University, which are full-scale with high-resolution; users have the sense of "being there", or the sense of presence experienced in the environment resulting from cognitive processes (Chan and Weng, 2005).

The sense of presence (see http://www.presence-research.org) is generated from human senses of sight, sound, taste, smell, and touch. In virtual environments, three conditions are required to generate the sense of presence through perception (Lombard and Ditton, 1997): image quality, image dimensions, and view distance. Studies found that high quality or resolution of images (Reeves et al., 1993), large scale or dimensions of images (Reeves et al., 1993), and the closer distance between viewers and images (Yuyama, 1982) generated greater sense of presence.

The immersive projection system fulfils these requirements, for it closely approximates actual size and distance with full-scale, high-resolution 3D objects generated in real time. For instance, C2 and C4 both are 12' by 12' spaces, in which the user is surrounded by three-dimensional images, projected in real time on three walls and the floor (currently, C2 is replaced by C4). C6 is a 10'x10'x10' room in which computer generated images are projected on all four walls, the ceiling, and the floor to deliver an enclosed, interactive, and fully immersive experience. Along with the high-end projectors, the system can produce up to 4096x4096 pixel images totaling over 16.7 million pixels per wall. Forty-eight dual-CPU workstations send images to 24 digital cinema projectors. Images generated by these projectors would have high resolution of approximately 1165 pixels per square inch (http://www.vrac.iastate.edu/facilities.php). This resolution gives users a clear and detailed display of a virtual environment.

Thus, C2, C4 and C6 could create vivid sense of presence, and experiments conducted inside are valid replications of the experiences derived from real-world interactions. Therefore, applying VR technology as a study tool to represent an environment and examining design through perception is very appropriate (Patel et al., 2002). Because visual perception provides us with more content and meaning than other senses, it more easily triggers the sense of presence.

## 6 Design in Virtual Environments

A Virtual Architectural Design Tool (VADeT) was created in the C2, which had a number of metaphorical icons (see Figure 1 left image) that served as design tools for generating, modifying, and editing three-dimensional objects of architectural elements (see Figure 1 right image). There were also tools for defining dimensions, materials (see Figure 2 left image), and colors (see Figure 2 right image) of the objects. By using these tools in the system, users could create a design in a synthetic VR facility (Cruz-Neira et al., 1993; Chan et al., 1999).

Through a number of kitchen designs accomplished (see Figures 5 and 6) in the VADeT system by architecture students compiled with protocol analysis (Ericsson and Simon, 1984), it was found from the protocol data that the overwhelming sense of immersion and projection in the Virtual Reality environment

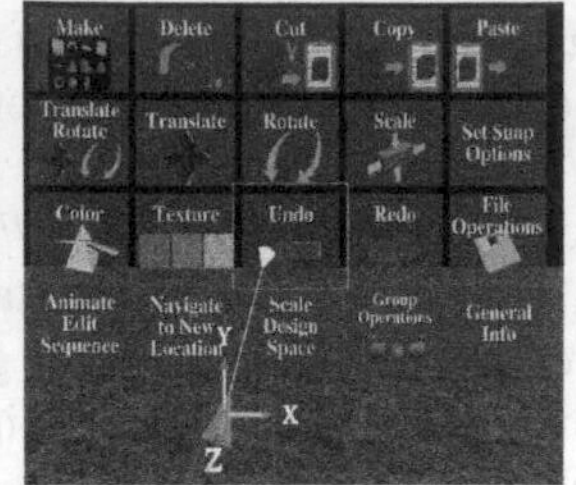


**Fig. 1.** Icons (left) and menus (right) used in the VADeT system.

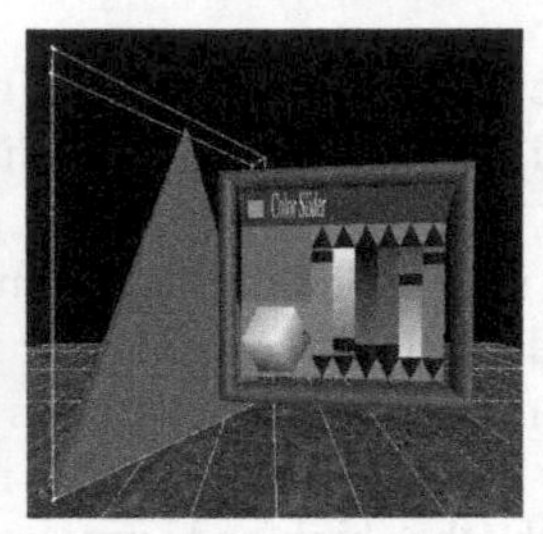


**Fig. 2.** Materials (left) and colors (right) used in the VADeT system.

has altered design behavior and thinking routine. There were several interesting findings in the experiments.

1. In conventional designs, designers would heavily rely on using scale to get accurate measurements for spatial layouts. In the full scale environment of this system, there were no physical scales available at the time of this experiment, so designers used their own bodies as the representation of scale for design.
2. Subjects focused much attention on the proportions of each object, their spatial relationships with adjacent objects, and their locations in the space that fulfilled functional links and visual connections with other objects in the scene. The representation of adjacency bound closely to geometric relationships between objects.
3. Design processes in the virtual environment are almost purely visual, with much attention devoted to the sizing, texturing, and coloring of the details of the objects; and less on reasoning and logical problem solving. The formation of representation is visually driven.
4. No alternative design solutions were created or considered in the experiments and the entire design process was linear progression. This might be due to the large amount of workload spent in the environment.

In short, the two subjects' design processes seemed more intuitive than deliberate. It could be the case that their perception was overwhelmed by 3D images and their thinking processes were driven mainly by geometric (visual) thinking instead of conventional logical reasoning (Chan et al., 1999; Chan, 2001b). It might also be

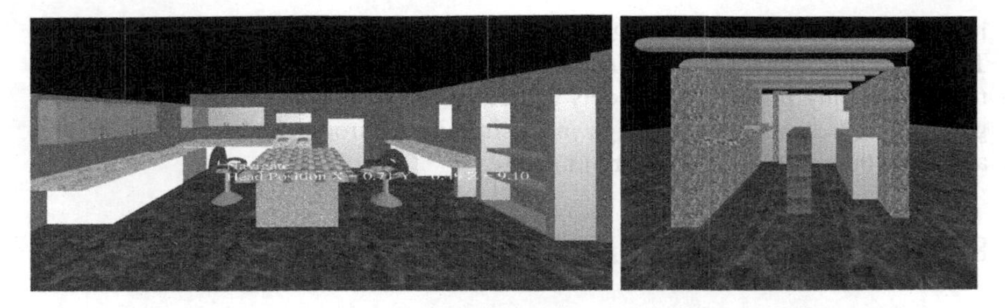

**Fig. 3.** Design examples A (left) and B (right) done in the VADeT system.

due to the fact that the full scale of the immersive virtual environment creates a very strong sense of presence (Chan and Weng, 2005). Therefore, designers would pay attention to the objects created and ignore overall functional layouts. As such, design cognition, under this new context, may be adjusted automatically to accommodate the new sensations created by this exotic visual world. This example explains the different design representation used in the C2 environments, which might also have changed the design strategies used for thinking. It also supports the notion that design could be characterized as a construction of representations from time to time (Visser, 2006).

## 7 Representation in Infinite Environments

The unlimited landscape and infinite scale of horizon in virtual environments have also generated an interesting new dimension to representation. The design of the virtual universe simulating the outer space in C6 is an interesting example (see the left image in Figure 4) for it is not associated with any real world problem solving experiences.

**Fig. 4.** The asteroid field shown in C6 (left), the Jupiter and Moons in the virtual universe (right). (Modeled by Christian Noon & Brandon Newendorp.)

The design representation used in this virtual universe relies heavily upon imagination for designing the asteroid, and its mental representation is arbitrary. This arbitrary nature is due to the fact that: (1) the sense of distance and the dimensions of the object in the huge space is relevant and not absolute (see the right image in Figure 4), (2) the design is not for solving realistic real world problems. Therefore, the internal problem representation used in this design is not linked to reality, and any images or concepts generated are purely imaginative. In this regard, the representation for design in the infinite environments is visually driven and everything created is arbitrary in nature.

## 8 Perception of the Virtual World in C6

One of the unique characters of the virtual space in C6 is its vivid full scale environment which gives viewers a fully immersive sense of presence (Chan and Weng, 2005). Perception in C6 is very convincing, allowing viewers to understand the impact of the environment through visual sensory data. Designers could gather information visually to identify cognitive impacts from the environments for evaluating design.

Studies on sensory impacts or casual effects in environments could be conducted in physical buildings to observe results first hand. However, it would be too costly and time-consuming to modify the setting if the task is complex and the settings are complicated. A VR environment, however, is suitable for such studies. Virtual environments that can generate a high degree of presence are thought to be more effective and better suited to task performance (Slater, 1999; Nunez and Blake, 2001). A series of experiments on simulating an office environment located on the seventh floor of the General Services Administration headquarters in Washington DC was conducted to study whether color, materials, or view would have big impact to human cognition.

In this series of studies, a virtual model was developed to realistically reflect the physical environment and served as the master model (see the left image in Figure 5), which was displayed in C6 (the right image in Figure 5). Afterwards, colors of the ceiling and walls were changed (Figure 6) to blue and red on the wall and ceiling in the first experiment. In the second experiment, materials of the furniture and partition walls were changed to oak, cherry and marble (Figure 7), and the window views were also changed from building images to four different natural landscapes (Figure 8) in the third experiments. Photos shown on the figures are for explanation purposes.

Thirty-one subjects from various colleges participated and viewed the master model first to serve as the visual reference, and then to justify the remaining models in C6 sequentially, and responded with the level of comfort score ranging from 0 to 9 for each model. Score with zero meant the subject found the change very unsatisfactory and nine was very satisfactory.

Results showed that blue wall color is the most welcome one by the subjects among the color group with highest average score of 5.9 (Chan, 2007a) in the first experiment. The oak material used for the furniture with average of 5.97 is the best one in the second experiment (Chan, 2007b). In the third experiment, the view to the outside world with an autumn scene has the highest average scores of 6.85 that suppress the perception of materials and colors.

**Fig. 5.** The master model of an office (left) and its display in the C6 (right).

**Fig. 6.** Different color of the model.

**Fig. 7.** Different material of the model.

**Fig. 8.** Different views on the windows.

# 9 Perception and Representation in Collaborative Artificial Spaces

Representation and perception have new dimensions in the collaborative artificial spaces which could be explained by the virtual worlds of Facebook and Second

Life. Second Life is a 3D virtual environment entirely built and owned by its residents (http://secondlife.com/). After it was opened to the public in 2003, it grew tremendously, with more than 100,000 people inhabited from around the world. Facebook is a social utility that connects people with their friends. The *New York Times* reported as Facebook will reach two million users in 2009 (Stone, 2009), it becomes the dominant social ecosystem and an essential personal and business networking tool to link people across space and time. The representations used in Facebook are old pictures or text of news articles.

In the Facebook space, a new sub-culture emerges due to the changes in: (1) users' communication method of posting, (2) different and new vocabularies used (abbreviation of words), (3) sense of individual identity expecting to completely see people but incompletely to be seen by other people, (4) different social interaction among virtual friends, and (5) psychological desire to share intimacies. These changes have created different recognition of identity and privacy in collaborative artificial spaces. People want to remain private and maintain separate social realms or a modicum of their privacy. As such, there is the tendency for individuals to associate only with like-minded people of similar age and ethnicity and a new social psychology of homophily is formed.

The same phenomena of identity and privacy issues have occurred in the Second Life environment. For instance, people create their own special and intended representations to show identifications, which could be the things (avatar) that they subconsciously intend to be but are not feasible in reality. People also present their "real" selves or expected shelters in certain details, which is the critical sense of realism and presence in the virtual artificial world. Such changes of representation would cut through arbitrary social barrier in the virtual space.

## 10 Conclusions

Design representation experiments have shown the subtle changes involved when designing in virtual environments. Furthermore, putting the representation and perception in a larger perspective and considering the new culture trend generated by the rapidly changing information technology, a new question emerges: if thinking has been impacted by a new culture, will design be changed accordingly? Affected by the new culture, the sense of privacy will seem either more open or conservative. Consequently, will the change in sense of privacy challenge the conventional sense and consequently require a new definition of private space? Regardless of whether the move is toward more or less privacy; the sense of self, one's identity, and the boundary of space in the virtual environments are rather arbitrary and reckless. Finally, would design require more ethics and loyalty? These issues should be further explored under the premises that design representation, changed by the high-tech development of media and evolution of culture has ultimately changed our design thinking and our spatial identification in the virtual environments.

## References

Akin, O.: Psychology of Architectural Design. Pion, London (1986)

Chan, C.S.: Cognitive processes in architectural design problem solving. Design Studies 11(2), 60–80 (1990)

Chan, C.S.: An examination of the forces that generate a style. Design Studies 22(4), 319–346 (2001a)

Chan, C.S.: Design in a full-scale immersive environment. In: Proceedings of the 2da. Conferencia Venezolana sobre Aplicaciones de Computadoras en Arquitectura, pp. 36–53 (2001b)

Chan, C.S.: Evaluating cognition in a work space virtually. In: CAADRIA 2007, pp. 451–458 (2007a)

Chan, C.S.: Does color have weaker impact on human cognition than material? In: CAAD Future 2007, pp. 373–384. Springer, Amsterdam (2007b)

Chan, C.S.: Design Cognition: Cognitive Science in Design. China Architecture & Building Press, Beijing (2008)

Chan, C.S., Hill, L., Cruz-Neira, C.: Can design be done in full-scale representation? In: Proceedings of the 4th Design Thinking Research Symposium – Design Representation, pp. 139–148. MIT, Boston (1999)

Chan, C.S., Weng, C.H.: How real is the sense of presence in a virtual environment? In: Bhatt, A. (ed.) Proceedings of the 10th International Conference on Computer Aided Architectural Design Research in Asia, TVB School of Habitat Studies, New Delhi, pp. 188–197 (2005)

Cruz-Neira, C., Sandin, D., DeFanti, T.: Surround-screen projection-based virtual reality: The design and implementation of the CAVE. In: Proceedings of the ACM SIGGRAPH 1993, pp. 135–142 (1993)

Eastman, C.: New directions in design cognition: studies of representation and recall. In: Eastman, C., McCracken, M., Newstetter, W. (eds.) Design Knowing and Learning: Cognition in Design Education, pp. 147–198. Elsevier, Amsterdam (2001)

Ericsson, K.A., Simon, H.A.: Protocol Analysis: Verbal Reports as Data. MIT Press, Cambridge (1984)

Fink, E.: Changes of Problem Representation: Theory and Experiments. Physica-Verlag, New York (2003)

Goldschmidt, G.: Design representation: private process, public image. In: Goldschmidt, G., Porter, W.L. (eds.) Design Representation, pp. 203–217. Springer, London (2004)

Hesse, M.: Models and Analogies in Science. University of Notre Dame Press, Indiana (1966)

Larkin, J.: The role of problem representation in physics. In: Gentner, D., Stevens, A. (eds.) Mental Models, pp. 75–98. Lawrence Erlbaum, Hillsdale (1983)

Lombard, M., Ditton, T.: At the heart of it all: The concept of presence. Journal of Computer-Mediated Communication 3(2) (1997), `http://jcmc.indiana.edu/vol3/issue2/lombard.html` (last accessed: March 2009)

Nunez, D., Blake, E.: Cognitive presence as an unified concept of virtual reality effectiveness. In: Chalmers, A., Lalioti, V. (eds.) Proceedings of the 1st International Conference on Computer Graphics, Virtual Reality and Visualization, pp. 115–118. ACM Press, New York (2001)

Patel, N.K., Campion, S.P., Fernando, T.: Evaluating the use of virtual reality as a tool for briefing clients in architecture. In: Williams, A.D. (ed.) Proceedings of the Sixth International conference on Information Visualization, pp. 657–663. IEEE, Los Alamitos (2002)

Reeves, B., Detenber, B., Steuer, J.: New televisions: The effects of big pictures and big sound on viewer responses to the screen. In: Information Systems Division of the International Communication Association, Washington, DC (1993)

Rowe, P.: Design Thinking. MIT Press, Cambridge (1987)

Schon, D.A.: The Reflective Practitioner. Temple-Smith, London (1983)

Simon, H.A.: The Science of the Artificial. MIT Press, Cambridge (1969)

Slater, M.: Measuring presence: A response to the Witmer and Singer presence questionnaire. Presence 8(5), 560–565 (1999)

Stone, B.: Is facebook growing up too fast? New York Time (2009), http://www.nytimes.com/2009/03/29/technology/internet/29face.html (last accessed: March 2009)

Visser, W.: Designing as construction of representations: A dynamic viewpoint in cognitive design research. Human-Computer Interaction 21, 103–152 (2006)

Yuyama, I.: Fundamental requirements for high-definition television systems. In: Fujio, T. (ed.) High definition television [NHK Technical Monograph]. NHK Advanced Broadcasting Systems Research Division, Tokyo (1982)

# Design Paradigms for the Enhancement of Presence in Virtual Environments

Rivka Oxman

Technion Israel Institute of Technology, Israel

**Abstract.** Virtual environments are being designed and implemented to accommodate diverse experiences of real space. Form and content are two basic concepts that have a significant impact on the sense of presence in virtual environments. Architectural design has a particular meaningful role in designing and creating form in which content takes place. Recent explorations found that these concepts may open a new understanding of innovative design paradigms that can induce the sense of presence. This chapter discusses current research in the design of presence in virtual environments. It presents and investigates the impact of two paradigms that integrate the components of form and content: the first is termed as "task-based design" and the second is termed as "scenario-based design".

**Keywords:** virtual environments, presence, Virtual Reality, virtual architecture, task-based design, scenario-based design.

## 1 Introduction

Virtual Environments (VEs) are currently being designed and implemented to accommodate diverse experiences of real space. Simulation of spatial reality has been a key role in duplicating the experience of real space. The immense growth in computer and Virtual Reality technology has resulted in the development of new approaches to the design of virtual environments. Current definitions of Virtual Environments are related to their performance as digital spaces. Virtual Environments are defined as computer programs that implement digital worlds with their own "physical" and "biological" laws" (Heudin, 1998); environment that contains objects and an interface linking the user and the environment (Sikora et al., 2000); a collection of objects, each object corresponding to things, space, and people (Gu and Maher, 2003). In digital spaces the form is based on simulation of consistent and logical transformations related to program and content.

There are two distinct types of virtual environments in which architectural design has a particular meaningful role in designing and creating the form in which content takes place: *Virtual Reality (VR)* and *Real Virtuality (RV)*. In *Virtual Reality,* form is based upon the simulation of environmental realism and content is the cognitive and instrumental performance enabled by the verisimilitude of the scene.

X. Wang & J.J.-H. Tsai (Eds.): Collaborative Design in Virtual Environments, ISCA 48, pp. 41–49.
springerlink.com

In *Real Virtuality,* form is based upon the generation of familiar scenes where content is based on typical associations and actions related to real life scenarios.

The domination of a virtual environment is related to immersion. In the most technically advanced applications of "immersive" VR, the user is essentially isolated from the outside world and fully enveloped within the computer-generated environment. There are various factors that are important in order to achieve immersion. For example, there are aspects related to form such as simulations of lighting, resolution, speed, and level of reality; there are also content-related aspects that are relevant to the task that the user is required to perform in a virtual environment. However, this kind of VR which simulates spatial and visual experience often reduces the cognitive space in which people can be active and creative in a way similar to everyday life.

In order to go beyond this concept, current research is looking for a richer set of concepts to enhance the richness and complexity of our experience in virtual environments. Among the challenging problems today is the achievement of a sense of presence in the virtual environments which might duplicate, replace, or improve the human sense of "being there". Research in the experience and evaluation of virtual environments has investigated significant concepts that can be applied, due to technological developments, in the design of virtual environments. Among them are the concepts of presence (Slater, 2002) and place (Mitchell, 1995; Kalay, 2001).

Given the centrality of these issues for VE's development, presence research is becoming a multi-disciplinary field of some importance. The role of architecture and design has particular significance in this emerging field. Being there – the experience of presence in mediated environments is associated with multidimensional perception and various cognitive processes. These subjects are of significance to the field of virtual design, since they provide a theoretical basis for understanding and developing novel design paradigms for the design of presence-rich virtual environments. Novel design paradigms should include and integrate both traditional design experience and the cognitive understanding of space, function and structure with emerging new concepts related to mediated environments.

This chapter reports on research to define, model and evaluate the architectural and design contributions to the strengthening of presence and, particularly, the role of human behavior in these environments.

## 2 Presence: Media-Form and Media-Content

Scientists in the field of psychology have developed the theory of a sense of presence (IJsselsteijn and Riva, 2003, Slater, 2002). According to them, the main feature of designing virtual environments is the sense of place, which contributes to the sense of presence. According to Wijnand and Riva (2003) there are various categories that affect the sense of presence. Among these characteristics, Slater (2002) has made a distinction between external characteristics associated with the media and internal characteristics associated with the user. Media effects are considered to be objective and well-defined while the user behaviour is considered to be subjective.

*Form* and *content* are basic concepts in understanding presence (Slater, 2002). Both are known to have a significant impact on the sense of presence. Various types of presence can be achieved in virtual environments according to the application of different types of the media-form and the media-content in the design of mediated environments. These will be explained below.

*Media-form* is defined by the physical properties of the display that enable the activation of virtual environments. The physical environment is responsive to the creation of multisensory stimuli that activates perception, cognition and emotion. For example, there are three categories that are associated with media form: sensory information, the level of control over the sensory mechanism, and the ability to modify the environment.

*Media-content* refers to objects, actors and other aspects of the environment that are represented by the medium and allow a flow of events known as the "narrative" or the "story". Factors of media content are responsible for keeping the user interested and involved.

Slater (2002) provided good examples of the achievement of interesting and engaging context in different settings by mixing principles of content and form. According to Slater, "being there" or "being present" means activating the perceptual, cognitive and mental systems in a way similar to real situations where human behaves as if he/she is there experiencing similar thoughts and actions. The sense of presence can be achieved even when the level of immersion is not high. What is important is the creation of rich and complex environments that could induce the human feeling of "being there". For example, in the design of a music hall, content can be conveyed by sensory and visual effects of the place such as sound, and the spatial experience can be achieved by visual and immersive effects. Both of form and content can create interesting and engaging environments for having a sense of presence. Furthermore, such a place can offer the social sense of place by providing opportunities to meet, recognize a face, and be surprised by meeting, or other forms of social contents.

This approach to the dual concepts of form and content open new directions to experiment with and develop new design paradigms that can induce an enhanced sense of presence in virtual environments. In the following section, we illustrate the exploitation of the concepts of form and content as the basis for the development of design paradigms for virtual environments.

## 3 Design Paradigms That Induce Presence

Designing a place in virtual environments should be rich enough to activate all the components of human perception, cognition and emotion. The goal of all presence and interaction technologies is to achieve a high level of presence coupled with a seamless functionality of interaction. There exists a base-line of important definitions in presence research that attempts to generalize the research issues above the level of specific technological applications.

Architectural design has a particular significance in this emerging design field, since it is a field that has been traditionally engaged with the creation of rich human experience through the design of the physical environment.

In our own research, we found the dual concepts of form and content are extremely useful. Presence appears to be affected by perceptual and cognitive factors related to the spatial-temporal-physical aspects of scenes that characterize both concepts.

This diverse and complex body of influences constitutes one pole of presence research related to architecture and design. We are currently exploring types of paradigms that integrate the two components of form and content together. Designing a place in virtual environments should be rich enough to activate all the components of the human perception, cognition and emotion. Architectural design has a particular significance in this emerging design field, since it is a field that has been traditionally engaged with the creation of rich human experience through the design of the physical environment. Furthermore, architectural design has historically developed a rich conceptual vocabulary for dealing with the description and evaluation of environmental forms.

We are now experimenting with three design paradigms in developmental and empirical situations: the first is termed as task-based design, the second is a scenario-based design and the third is termed a performance-based design. Task based design is implemented as a Virtual Reality technology and was presented elsewhere (Oxman et al., 2004), the other two paradigms are related to "Real Virtuality" and illustrate these ideas (Oxman et al., 2004; Kolomiski, 2007). The former two paradigms that integrate form and content are currently under exploration. Both of them are elaborated in the following sections.

## 4 Task-Based Design

In our approach to task-based design, the form is based upon the simulation of environmental realism and the content is the cognitive and instrumental performance enabled by the verisimilitude of the scene. Slater (2002) suggested that presence in virtual environments includes the following three aspects: a sense of "being there", domination of the virtual environment over the real world, and activation of the participant's memory as if he is in an actual location rather than a compilation of computer-generated images and sounds.

Task involvement, the "control" of visual scene and scene transitions by the participant (usually the participant's movements) are important issues in the design of such task-based environments. The way users can interact with displayed images, move and manipulate virtual objects, and perform other actions in a way that engenders a feeling of actual presence, and immersion of their visual sense, in the simulated scene is one of them (Wijnand and Riva, 2003).

HabiTest (Palmon et. al., 2004) is a "task-based design" environment (in some way) which is employed for the development of virtual living environment for home modification processes. One of the major challenges facing the design of home modification environments is to succeed in adapting the environments in a way that enables an optimal fit between the individual and the setting in which he or she operates.

According to Nash et. al. (Nash, 2001) under certain conditions when a task is more meaningful, interesting or competitive to the user, the level of presence is generally improved, even in the absence of high immersion. According to Palmon et. al. (2004) the goals in designing the HabiTest interactive model and the tool (termed "LES" - Living Environment System) were to display three-dimensional renderings of specific environments, which respond to user-driven manipulations such as navigation within the environment and alteration of its design.

According to Palmon et. al. (2004) the main purpose of the HabiTest model was to test whether the environment is an accessible environment, to test and to see where the barriers are. The virtual environment was being developed for testing various everyday activities within a context of a domestic spatial environment. The assumption was that the accessibility of a real environment through the performance of humans in a virtual space is based upon the approach that "place understanding" in virtual environments may be similar to a place in real environments. Thus, humans may perform in VEs in a similar way to the way they perform in real environments. For example, it was found by Darken and Silbert (1996) that the same principles of environmental design that are implemented to assist way-finding in the physical world (e.g., directional cues) help way-finding in virtual worlds.

The LES enabled the development of an interactive environment that can be used to test users' abilities to identify and modify accessibility barriers while navigating within it. The construction and simulation of these environments were carried out using EON Reality's (www.eonreality.com) tools. These tools enabled a rapid development of interactive 3D environments that were easy to navigate in real time while performing accurate collision detection. Accurate collision detection (which was available primarily in mechanical and non-interactive simulations) enhanced our ability to gather relevant data from the simulation process. In previous generations of VR tools, the collision detection was limited to a bounding box. Such bounding box was a rough approximation of the user's body contours and left out many of the fine details such as curves, gaps and protrusions, which are necessary to accurately represent the body. With the EON Reality platform, we could not only identify each collision and record the occurrences into a database, but also give auditory, visual and haptic feedback to the user in order to

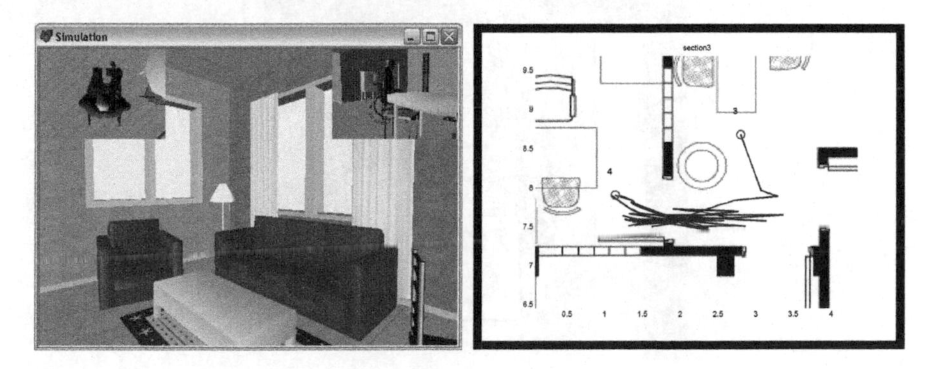


**Fig. 1.** Usability test and evaluation in task-based design

avoid physically invalid positions. The user could not navigate to position where the wheelchair or a part of his or her body overlaps with another object (e.g., a wall, a door, a table or its legs, etc).

The validity of this simulation allowed the user to navigate in the VR in a way that is similar to the navigation in the real environment. The user could not only identify barriers in his/her way but even identify corners or narrow passages which, although passable, would be difficult and inconvenient to navigate on a daily basis due to the number of movement and necessary collisions.

In general, the task-based design approach is related to both form and content. Form is achieved by simulating a specific environment and the content is achieved by providing tools for testing performance of a set of specific functions and tasks.

## 5 Scenario-Based Design

Our approach to "scenario-based design" attempts to achieve a sense of spatial reality in virtual environments through the materiality, and the mobility of the subject(s). One of the most interesting challenges of VEs is the achievement of the complex characteristics of *real-time interaction in social situations*. This represents another level of presence enhancement in VEs that involves various aspects of socialization. These include real-time social action-reaction phenomena in visual-temporal space, communication-response, etc.

**Fig. 2.** Co-presence in scenario-based design

In the following particular application, the utilization of the concepts of "content" and "form" are elaborated and tested in a situation of virtual co-presence. "Co-presence" (Lombard and Ditton, 1997) is created by the integration and the interaction of the two types of presence, physical presence (form) and the social presence (content). Scenario-based design creates choice and opportunities for social interaction simultaneously in various virtual places. Figure 2 illustrates an example of a meeting between a visitor and a painter in a virtual museum.

The unique quality of this environment is that the painter can meet the visitor in the museum upon her/his request, or alternatively invite the visitor to his/her studio. This paradigm supports a real or imaginary scenario that can be associated with places such as museums, schools etc. The meeting creates the quality of co-presence. In addition, the sense of presence is enhanced by the unique experience provided by the virtual, an experience possible, but usually beyond, that of real environmental experience in such environments.

This application characterizes a *user-centered augmented service environment* which attempts to enhance the real-time accessibility and control of a range of functions that are integrated with the visual scenes. In such expanded service Augmented Reality environments the role of architecture and design factors satisfies a range of important cognitive requirements.

## 6 Evaluation Criteria in Presence Research

According to Oxman et al. (2004), evaluation criteria are the main challenge in research related to presence. Defining the link between the presence measures and achieving good design and the way we can determine the extent to which a user feels present in a virtual environment are the most significant issues in the evaluation of how successful is the design of a virtual environment.

There are various methods commonly used for measuring "presence" such as self assessment, behavioral measures and physiological variable that examine actions and manners exhibited by the users in response to objects or events, etc. (Insko, 2003). However, in order to address the complexity of the interpretation of presence as it relates to the design of virtual places, evaluation issues that are associated with the design and creation of virtual environments are most relevant. In the design domain of large scale architectural virtual environments, Knight et al. (2003) in their work on the construction of a virtual university campus discussed and presented issues related to the usage of naturalistic interface in testing the experience of presence in VR environments.

Since there are various design paradigms for virtual environments that have different objectives, as we have presented and demonstrated in this work (task-based design and scenario-based design) and since a good design can add to the success of achieving the sense of presence that users sense, there should be a particular way to evaluate presence in each approach. Good design is a design that allows users the type of interactions of a specific environment.

## 7 Conclusions

In exploring the suitability of new design paradigms for the construction of virtual places, we found that the concepts of form and content stimulate insightful innovation with respect to the potentials of the design of virtual environments. In design, they offer a good point of departure for more experimentally-driven design approaches for virtual environments as well as for imaginative exploitation of the virtual in order to create a sense of the hyper-real.

In order to advance such developmental hypotheses of new paradigms of design, the evaluation and measurement of the sense of presence in designed mediated environments have become essential. Since there is no well-defined and universally-accepted methodology for the evaluation of design and for the characterization of good design, there is a need to address the complexity of the interpretation of *"presence in place"* as it relates to design of such places. Evaluation issues that are associated with the design and creation of virtual environments have become an important research priority.

## Acknowledgement

The work presented in this chapter is based on the extension of prior work presented in previous papers (Oxman et. al., 2004; Palmpn et. al., 2004, Oxman, 2004). The work on immersive environments for disabled users is the Ph.D. research work carried out by Orit Palmon under the supervision of Prof. Patrice L. (Tamar) Weiss of the University of Haifa, Department of Occupational Therapy, and Prof. Rivka Oxman of the Technion, Faculty of Architecture and Town Planning. The work on scenario-based design is based on the Ph.D. research work carried out by Rina Kolomisky under the supervision of Professor Rivka Oxman.

## References

Gu, N., Maher, M.L.: A grammar for the dynamic design of virtual architecture using rational agents. International Journal of Architectural Computing 4(1), 489–501 (2003)

Darken, R., Silbert, J.: Wayfinding strategies and behaviors in large virtual worlds. In: Proceedings of ACM CHI 1996: Human Factors in Computing Systems, pp. 129–142 (1996)

Heudin, J.-C. (ed.): VW 1998. LNCS (LNAI), vol. 1434. Springer, Heidelberg (1998)

Insko, B.E.: Measuring presence: subjective, behavioral and physiological methods. In: Riva, G., Davide, F., IJsselsteijn, W.A. (eds.) Being There: Concepts, Effects and Measurement of User Presence in Synthetic Environments. IOS Press, Amsterdam (2003)

Lombard, M., Ditton, T.: At the heart of it all: The concept of presence. Journal of Computer-Mediated Communication 3(2) (1997)

Kalay, Y.E., Marx, J.: Architecture and the internet: Designing places in cyberspace, Working Paper, University of California, Berkeley (2001)

Knight, M., Brown, A.G.P., Hannibal, C., Noyelle, C., Steer, O.: Measurement of presence in Large Scale virtual environments. In: Proceedings of ECAADE 2004, Austria, Graz (2004)

Kolomiski, R.: Designing presence in the virtual museum: A scenario based approach, MSc Thesis, Faculty of Architecture and Town Planning, Technion, Haifa (2007)

Mitchell, W.J.: City of Bits, Space, Places and the Infobahn. MIT Press, Cambridge (1995)

Nash, E.B., Edwards, G.W., Thompson, J.A., Barfield, W.: A review of presence and performance in virtual environments. International Journal of Human-Computer Interaction 12, 1–41 (2000)

Oxman, R., Palmon, O., Shahar, M., Weiss, P.T.: Beyond the reality syndrome: Designing Presence in Virtual Environments. In: Proceedings of ECAADE 2004, Copenhagen, pp. 35–42 (2004)

Oxman, R.: Design paradigms for the enhancement of presence in virtual environments: Theory, development and evaluation. In: Maher, M.L. (ed.) DCC workshop on Virtual worlds, pp. 33–38. MIT, Boston (2004)

Palmon, O., Oxman, R., Shahar, M., Weiss, P.T.: Palmon, O, Oxman, R, Shahar, M and Weiss, PT: 2004, Virtual environments as an aid to the design and evaluation of home and work settings for people with physical disabilities. In: International Conference Series on Disability, Virtual Reality and Associated Technologies, New College, Oxford (2004)

Slater, M.: Presence: Teleoperators and Virtual Environments. MIT Press, Cambridge (2002)

Sikora, S., Steinberg, D., Lattaud, C.: Integration of Simulation Tools in On-Line Virtual Worlds. In: Heudin, J.-C. (ed.) VW 2000. LNCS (LNAI), vol. 1834, p. 32. Springer, Heidelberg (2000)

IJsselsteijn, W., Riva, G.: Being there: the experience of presence in mediated environments. In: Riva, G., Davide, F., IJsselsteijn, W. (eds.) Being There: Concepts, Effects and Measurements of User Presence in Synthetic Environments. IOS Press, Amsterdam (2003)

# Co-presence in Mixed Reality-Mediated Collaborative Design Space

Xiangyu Wang and Rui Wang

The University of New South Wales, Australia

**Abstract.** Co-presence has been considered as a very critical factor in shared virtual environments (SVEs). It is believed that by increasing the level of co-presence, the collaborative design performance in a shared virtual environment could be improved. The aim of this chapter is to reflect on the concept and characteristics of co-presence, by considering how Mixed Reality (MR)-mediated collaborative virtual environments could be specified, and therefore to provide distributed designers with a more effective design environment that improves the sense of "being there" and "being together".

**Keywords:** mixed reality, co-presence, shared virtual environments (SVEs), collaborative design.

## 1 Introduction

It is now widely believed that collaboration could add values to individual work in many aspects. As a type of user experience, the feeling of "being there", or called *presence*, is actually independent on any specific type of technology. It is the product of mind. However, with the improvement of immersive displays, computing and network technologies, more accurate reproductions and simulations of reality could be created. This makes people increasingly aware of the relevance and importance of the presence experience. The concept of presence has become an important research topic in such areas as cinematic displays, virtual environments, telecommunication and collaboration.

*Co-presence*, a sense of being together in a shared space, is also a critical factor of remote collaborative work within a shared environment. Co-presence consists of two dimensions: co-presence as mode of being with others, and co-presence as sense of being with others (Milgram, 1994). Mode of co-presence refers to the objective physical conditions which structure human interaction; while sense of co-presence refers to the subjective experience of being with others that an individual acquires within interaction (Milgram, 1994) Collaborative Virtual Environments (CVEs) should provide high level of co-presence which could encourage effective collaboration and communication between distributed users.

Lombard and Ditton (Lombard, 1997) reviewed a broad body of literature related to presence and have indentified six different conceptualizations of presence:

X. Wang & J.J.-H. Tsai (Eds.): Collaborative Design in Virtual Environments, ISCA 48, pp. 51–64.
springerlink.com 

realism, immersion, transportation, social richness, social actor within medium, and medium as social actor. Based on the commonalities between those different conceptualizations, IJsselsteijn et al. (Ijsselsteijn, Freeman, & De Ridder, 2001) suggested that those conceptualizations Lombard and Ditton identified can be roughly divided into two broad categories – physical and social. Physical presence refers to the sense of being physically located in mediated space, whereas social presence refers to the feeling of being together, of social interaction with a virtual or remotely located communication partner (IJsselsteijn et al., 2001). At the intersection of these two categories, co-presence could be identified as combining significant characteristics of both physical and social presence. Figure 1 illustrates their relationship with a number of media examples that support the different types of presence to a varying extent. For instance, Virtual Reality (VR) technologies could encourage relatively higher level of physical presence than a plain painting does; video conferencing may bring better sense of "being together" to users than email does, because it includes both physical presence and social presence.

It is apparent that social and physical presences are distinct categories and have some overlapping part to each other. The obvious difference is the role of communication, which is a major part of social presence, but not necessary to establish a sense of physical presence. In particular, a medium can provide a high level of physical presence without having the capacity for transmitting reciprocal communicative signals at all (Ijsselsteijn, de Ridder, Freeman, & Avons, 2000). In the other hand, people can experience a certain amount of social presence by using applications, which only provide minimal physical representation. Text-based or verbal-based online chatting could be examples under this condition. However, the two categories are still related to each other. There could be a number of common determinants, for instance, the latency during interaction is relevant to both social and physical presence (Ijsselsteijn et al., 2000). As Ijsselsteijn (IJsselsteijn, 2001) illustrated in Figure 1, applications such as videoconferencing or shared virtual environments are in fact based on providing a mix of both the physical and social

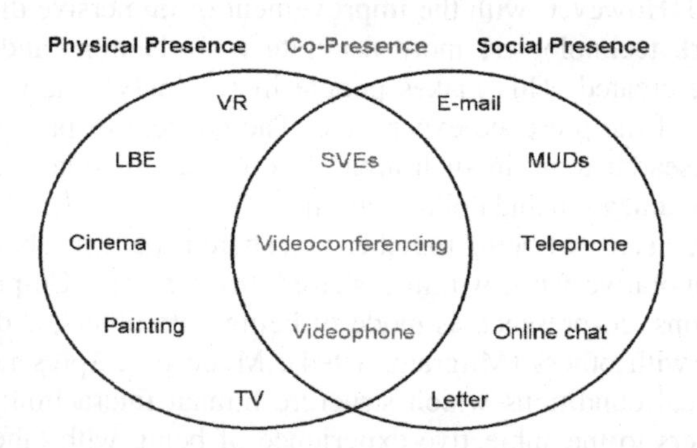


**Fig. 1.** A graphical illustration of the relationship between physical presence, social presence and co-presence, with various media examples. Abbreviations: VR = Virtual Reality; LBE = Location-Based Entertainment; SVEs = Shared Virtual Environments; MUDs = Multi-User Dungeons (Ijsselsteijn et al., 2001)

components. Therefore, in order to increase the level of co-presence, it is necessary to analyze the factors of both physical presence and social presence and enhance both of them together.

## 2 Co-presence Factors

This section introduces co-presence factors from literature review, which could be roughly divided into two categories: physical presence factors and social presence factors.

### 2.1 Physical Presence Factors

This section reviews and discusses several critical factors of physical presence that are closely related to virtual environments, including sensory inputs and outputs, visual display characteristics and other stimulus.

#### 2.1.1 Number and Consistency of Sensory Outputs

It is generally believed that the greater the number of human senses for which a medium provides stimulation, for instance, media sensory outputs, the greater the capability of the medium could produce a sense of presence (Anderson & Casey, 1997). For example, the media that provides both audio and visual stimuli could produce a greater sense of presence than that provide only audio or video. Short et al.'s study (Short, 1976) showed that greater social presence was reported by subjects after an audio-visual task-based interaction than an audio only one. Held and Durlach (Held, 1992) claimed that not only the number of sensory output channels an important factor in generating a sense of presence, the consistency of information in the different modalities is also a key issue. Failure to meet this criterion could result in unnatural and artificial user experiences.

#### 2.1.2 Image Quality

Lombard concluded that "the perceived quality of an image depends on many characteristics, including resolution, color accuracy, convergence, sharpness, brightness, contrast, and the absence of "ghosts" or other noise" (Lombard, 1997). Neuman's research (Neuman, 1990) showed that very high resolution images could evoke more self-reported presence than standard resolution images. Furthermore, Bocker and Muhlbach (1993) found that higher resolution images in a video conferencing system elicited reports of greater "communicative" presence. Images that are more photorealistic, for example a live-action scene or a photograph rather than an animated scene or a drawing, are likely to provoke a greater sense of presence as well (Heeter, 1992a). However, current virtual reality technology has not yet achieved a photo realistic appearance.

The size of a visual image has received greatest attention from researchers' concern about presence among all of the formal features. Larger images have been shown to evoke a variety of more intense presence-related responses. The research work by Reeves et al. (Reeves, 1993) showed that subjects who watched on a 70 inch screen (measured diagonally) reported significantly greater agreement with the statement "I felt like I was a part of the action" than subjects who watched on a 35 inch screen. Lombard et al. (Lombard, 1995) showed subjects 10 different short scenes featuring this same type of rapid point-of-view movement on a consumer-model television set with either a 46 inch or a 12 inch screen. Subjects who watched the larger images were more aroused (a skin conductance measure) and reported a greater "sense of movement," "enjoyment of this sense of movement," "sense of participation," and "involvement." At the same year, Lombart et al. (Lombard, Reich, Grabe, Bracken, & Ditton, 2000) used a screen size manipulation to show that viewers respond to social cues they encounter in nonmediated communication, such as apparent interpersonal distance, in mediated experiences including television viewing. The results showed that when the subjects watched attractive and professional news anchors deliver stories on a large (42 inch) screen they reported more positive emotional responses to anchors and to the viewing environment, and then selected a viewing position that represented a smaller withdrawal from the encounter, than when the people appeared on smaller (26 inch or 10 inch) screens (Lombard et al., 2000).

### 2.1.3 Image Size and Viewing Distance

It seems logical to expect that when people are physically closer to an image, they feel a greater sense of being a part of the image and therefore a greater sense of presence could be evoke (Lombard & Ditton, 1997). However, these two variables also act together to determine the value of a third variable, the proportion of the user's visual field that the image occupies, also known as viewing angle (Hatada, Sakata, & Kusaka, 1980) and field of view (Biocca, 1995). A large image and a large viewing distance, for instance in an IMAX theater, can result in the same proportion of visual field as a small image and a small viewing distance such as in a Virtual Reality head-mounted display (HMD). Some researchers suggest that a "sensation of reality" is stronger in the former configuration, however, this claim still needs more research findings to support.

Since the image size and viewing distance could act together on the level of physical presence, the use of egocentric such as HMD might be a good solution instead of projector with large screen as mentioned in the previous factor. A webcam could be attached onto the HMD and the virtual scene changes when the user changes his/her view point. However, due to the technique limitation, the HMD usually have lower resolution than computer displays do. To balance those variables, experiments should be carried out.

### 2.1.4 Motion and Color

It seems reasonable to conclude that moving images that provide the illusion of continuous action can more easily evoke presence than still images (Anderson, 1993). Also, it is assumed that color images should evoke a higher level of presence than those in black and white (Lombard & Ditton, 1997). Although it is lack of research in this area currently, this feature is worth to be considered in related system development.

### 2.1.5 Dimensionality

Traditionally, there are several ways to make two-dimensional (2D) images appear to contain a third dimension of depth. The 2D to 3D techniques are increasingly applied by designers of virtual environments (Heeter, 1992a), computer interfaces, and television graphics (Olds, 1990) to create the illusion that mediated objects have depth. The reason that three-dimensional techniques are considered so much is that three-dimensional visualization could evoke higher level of presence. Therefore 3D techniques should be imported where appropriate.

### 2.1.6 Aural Presentation Characteristics

Studies (Kramer, 1995) have shown that mediated sounds clearly are important in generating presence. The two most frequently identified characteristics of aural presentations in the context of presence are quality and dimensionality (Lombard & Ditton, 1997). Reeves et al. (Reeves, 1993) showed subjects scenes from action films and varied the fidelity of the soundtracks by controlling both frequency range (with a graphic equalizer) and signal to noise ratio (via tape "hiss" added to the recordings). The presentations with high fidelity sound were judged more "realistic," but it was the low fidelity sounds that made subjects feel more "part of the action". Therefore, they concluded that high quality audio is more likely to generate presence than low quality sounds; the scant available evidence is mixed (Reeves, 1993).

Biocca and Delaney suggested that (Biocca, 1995) because people hear in "three dimensions," the spatial characteristics of sound should be important for a sense of presence. Spatialization, or 3D sound, is an attempt to add these spatial characteristics to mediated sounds (Kramer, 1995). The volume (loudness) of mediated audio stimuli also may have an impact on presence, with particularly low and perhaps particularly high levels less effective than moderate ("realistic") levels (Everest, 1987).

### 2.1.7 Interactivity

The concept of interactivity is complex and multi-dimensional, but in this context an interactive medium is one in which the user can influence the form and/or content of the mediated presentation or experience as defined by Steuer (Steuer, 1995). The degree to which a medium can be said to be interactive depends on a number of subsidiary variables.

- The number of inputs: The number of inputs from the user that the medium accepts and to which it responds is an important variable of the level of interactivity. It could include voice/audio input, such as speech recognition systems which allow the computers to receive and make response to the voice commends, or TTA (text to audio) technology (Sallnäs, 2002), which converts text-based messages to audio; and haptic input such as keyboard, mouse or other touchable devices; gesture input such as body movements and orientation input by the help of suits with sensors or data gloves; and also other types of inputs, for instance, facial expressions, eye movements and psychophysiological inputs (Biocca, 1995) such as heart rate, blood pressure, body temperature, or even brain waves.

- Number (and type) of characteristics that can be modified by the user: The number (and type) of characteristics of the mediated presentation or experience that can be modified by the user determines the degree of interactivity (Lombard & Ditton, 1997). Steuer (Steuer, 1995) identifies the dimensions of temporal ordering (order of events within a presentation), spatial organization (placement of objects), intensity (of volume, brightness, color, etc.), and frequency characteristics (timbre, color). Others might include size, duration, and pace. Heeter (Heeter, 1992a) suggests that a highly responsive virtual environment, one in which many user actions provoke even unnatural responses could evoke a greater sense of presence than less responsive environments.
- The range or amount of change possible in each characteristic: Lombard and Ditton stated that the range or amount of change possible in each characteristic of the mediated presentation or experience is an important variable as well (Lombard & Ditton, 1997). They suggested that expanding the degree to which users can control each attribute of the mediated experience, and therefore the level of presence could be increased could enhance interactivity. For instance, in a virtual environment where a high level interactivity is enabled, where users can experience a three-dimensional world rather than two-dimensional ones, control volume of sound, move to wherever they want to and may even edit on the virtual environment.
- The degree of correspondence between the type of user input and the type of medium response: An important variable for interactivity and presence is the degree of correspondence between the type of user input and the type of medium response. (Steuer, 1995) suggests that the "mapping" between these two can vary from being arbitrary to natural.
- The latency: The latency, which reflects the speed with which the medium responds to user inputs is another variable. The ideal interactive medium responds in "real time" to user input (Lombard & Ditton, 1997). For example, video conferencing is highly interactive in terms of this criterion because interactions via video conferencing seem to occur in real time. On the other hand, the latency caused by computer hardware or bandwidth in virtual reality system could result in noticeable lag.
- Stimuli for other senses: Visual and aural stimuli may be the most common sensory outputs available in mediated experiences, but there are at least four others, each of which is likely to enhance presence: olfactory output (Biocca, 1995), body movement, tactile stimuli (Biocca & Levy, 1995), and force feedback (Heeter, 1992b). Adding the smells of food, flowers, or the air at a beach or in a rain forest to the corresponding images and sounds seems likely to enhance a sense of presence for media users (Hellig, 1992).
- Obtrusiveness: For an illusion of nonmediation to be effective, the medium should not be obvious or obtrusive -- it should not draw attention to itself and remind the media user that she/he is having a mediated experience. In 1992 Held and Durlach (Held, 1992) argued that presence requires a virtual environment to be "free from artifactual stimuli that signal the existence of the display". When possible, the user should not see edges of displays, speakers, microphones, measurement devices, keyboards, controls, or lights. This idea

applies to any medium; for example, Mitsubishi boasts that its television sets are "invisible except in brilliant sound and picture" (Held, 1992). Glitches or malfunctions in the operation of the medium (e.g., computer malfunctions, projection problems in a movie theater) make the mediated nature of the experience obvious and interfere with presence. Medium-specific formal features such as the use of text to identify news anchors and graphic logos to identify channels or networks also draw attention to the artificial and mediated nature of the presentation. Kim (Kim, 1996) suggested that noise, broadly defined as "information that is irrelevant to the intended communication regardless of the sensory channel through which it is transmitted" discourages presence. The form of a media presentation/experience can encourage or discourage noise (and presence) in a number of ways: a virtual reality system can be set up in a quiet room or a noisy arcade, the operator of a movie theater can take steps to discourage patrons from talking during the film, a family can watch television with bright or dim ambient light. The adoption of HMDs and 3D sound, as well as the natural interactivity in MR-Collab could generate the immersive sense for users to increase the level of presence.

- Live versus recorded or constructed Medium: It could be experienced that a mediated event as it happens at current moment, as the event occurred at an earlier time, or as it never occurred (Lombard & Ditton, 1997). The knowledge that a mediated event has been recorded or constructed may make it more difficult for users to perceive the experience as non-mediated. In MR-Collab, live video conferencing of customer and real-time synchronization of designers' behaviors are adopted to enhance the sense of presence.

## 2.2 Social Presence Factors

This sub-section discusses factors of social presence, including number of people and social realism that exists in an environment.

### 2.2.1 Number of People

One important social presence factor that may encourage a sense of presence is the number of people the user can (or must) encounter while using the medium. Heeter (Heeter, 1992a) suggests that "people want connection with other people more than any other experience. A medium that allows, or requires in the case of the telephone, the user to interact with at least one other person may evoke a higher level of presence than others. The ability to interact with larger numbers of people, for example, through multi-player virtual reality systems, video conferencing systems designed for interaction among large groups of people, or telephone conference calls may lead to even greater presence (Biocca, 1995; Steuer, 1995). MUDs is a well-known virtual environment system that includes multiple players; although it does not even have a graphic interface, a large number of people are deeply involved into this virtual environment (Bartle, 1990).

### 2.2.2 Social Realism

Lombart stated that (Lombard & Ditton, 1997) anyone who watches movies or television knows, the storylines, characters, and acting in some media content is more realistic than in others. A number of different labels have been used to identify this concept, including social realism (Dorr, Graves, & Phelps, 1980), a component of perceived realism (Potter, 1988), verisimilitude (Barker, 1988), plausibility (Elliott, 1983), and authenticity and believability. Social realism is distinct from perceptual realism, which is a characteristic of media form rather than media content. While social realism is usually applied to traditional media content, a virtual world can also contain more or less social realism (Lombard & Ditton, 1997). It is obvious that a world with scenes that largely differ from users' experiences from daily life, for instance, people with green skin or moon in a triangle shape is less likely to evoke presence.

# 3 Case Studies

This section introduces three systems that adopt Mixed Reality, Augmented Reality or other network technologies to increase the level of co-presence to benefit people with collaborative design tasks:

## 3.1 MR-Collab

MR-Collab is a Mixed Reality-mediated Collaborative design system, which combines Mixed Reality, Augmented Reality (AR) and various channels of communication technologies. The entire system is physically distributed in three geographically separated rooms and two types of end clients are included: two distributed designers, and the customer. This system could seamlessly connect distributed spaces together and make users located in different places feel as they are co-present in the same working environment with the support of Mixed-Reality technologies. Figure 2 shows the working environment of MR-Collab.

In the MR-Collab system, multiple channels of sensory outputs are adopted including both visual and audio features. Furthermore, there are a number of visual perceptual environments/stimuli such as virtual objects, virtual avatar, video conferencing images etc. Those different visual stimuli could affect on each other as well. For instance, the feature of virtual avatar and video conferencing images could work together to represent a designer, which could evoke higher level of presence than video conferencing only. The consistency of information is also highly considered when designing the system to generate a seamlessly connected environment such as the combination of virtual avatar and video conferencing (consistency between different visual stimuli), or the combination of video images and verbal chatting (consistency between different outputs).

In the system, images with high level of resolution and photorealism are adopted instead of low resolution or cartoon-styled images to generate higher level of physical presence. Considering this factor, real-scaled virtual creatures and avatars are adopted rather than down-scaled virtual objects or avatars such as in Second Life environment. Continuous action and images rich in colors are adopted to evoke more presence. The virtual scenes in the distributed physical spaces synchronize with each other in real-time.

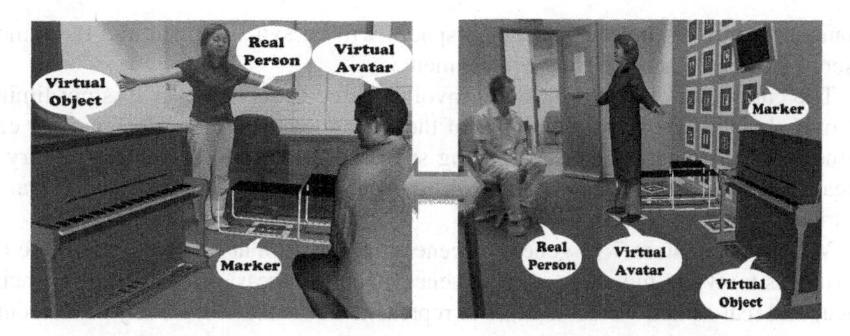


**Fig. 2.** Merging of real and virtual environments (Wang & Wang, 2008)

In MR-Collab, three dimensional virtual creatures and avatars are adopted. Rather than having three-dimensional looks in a two-dimensional screen, those virtual creatures and avatars are actually taken into the three-dimensional Mixed-Reality environment. People walk around those 3D virtual objects or avatars in their local physical environments. This feature could generate greater sense of "being there" than those virtual environments that show the entire scene on two-dimensional computer displays. At the mean while, three-dimensional sounds with spatial information could be adopted to evoke higher level of presence. For instance, the voice of the remote participants should contain their location information within the shared environment. Therefore a designer could be aware of where the other designer is working at through the direction of voices, even when the other designer is standing behind him/her and could not be seen by him/her visually.

In MR-Collab, interactivity between designers and environment or between designers is natural and tangible. Rather than using keyboard or mouse, users will use natural body gestures to control events as they do in physical environments. By this means users might not be reminded that they are "sitting in front of computers" but encouraged as they are actually in the shared environment.

Various inputs are adopted in MR-Collab systems. Several different trackers are used to capture users' appearances, voices and behaviors. Video cameras, microphones and several sensors are integrated in the system. Those different inputs will be discussed in the section of technical details.

In MR-Collab system, when designers are collaborating on a design task, they are able to edit both themselves' and others' creatures on the location, size, color and rotation variables. They are also able to create and make changes on their own virtual avatars to make them more recognizable to other users.

The MR-Collab system includes many features mentioned above. Users could look out in any directions; they could pick up, feel or move each virtual creature and change their variables and change the volume level of ambient sounds. Users are able to make whatever body movement as they could do physically. Within the system, natural gestures such as grabbing virtual objects, waving hands or turning heads are used as user inputs and the responses that users could get from system are also natural: they could see each other grabbing virtual objects, waving hands or turning heads. The system will make responds to user inputs in a real-time manner. Virtual scenes in distributed spaces will be synchronized and the remote users will share a same virtual environment.

The current model of MR-Collab involves three users however it is not limited to only three people. The scalability of the system enables more designers or customers be added into either in existing spaces or new spaces when necessary to create a larger environment. However, even involving more people, the framework of system is similar.

MR-Collab system will generate scenes that are similar to users' daily life experiences to evoke more sense of presence: facial expression (video conferencing face) and real-scaled virtual avatars to represent users' real-time body movements.

## 3.2 Virtualized Reality-Integrated Telepresence System

For remote collaboration in virtual environment, one conceptualized Mixed-Presence Groupware system was presented with the emphasis on two concerns. One is the spatial faithfulness, in another word, to provide means by which geographically dispersed users could perceive the environment and the space as if they are face-to-face. The other emphasis is the natural interaction support, such as gaze, gesture, and object manipulation. Technically, this system consists of two components, one tele-presence component and one tabletop component. The tele-presence component utilizes multiple cameras and full-scale displays to transmit and realize user's visual representation. The idea of this setup is to simulate communication in face-to-face environment, where each user has his/her unique perspective of view of the working environment. Although the contents of the working environment could be similar, the impression and perception in one's brain might vary dramatically. Careful attention needs to be paid when deciding what and how should each user see others within the virtual environment. It then leads to the question of the locations of the cameras and displays. In this system, the locations of the cameras are chosen with respect to the positions of the eyes of remote users. They are placed on top of the full-scale displays, as close to the "eyes" portion on the screen as possible. In addition to the tele-presence component, a tabletop component is aligned to promote the naturalism of the shared virtual environment. It is a round tabletop so that each user could be evenly seated around it. The seating position is consistent across all remote sites. Every user should agree that one is not allowed to take a seat, which has been taken by either a local user or a remote user. Otherwise, collision could happen, which interferes the sense of presence. The tabletop component also allows multiple, simultaneous manipulation via natural gesture instructions. Users are able to move and rotate the virtual objects, as they would do with real objects in face-to-face environment. In addition to that, some novel and advanced interactivities, which are not generally feasible in real world, can be applied on these virtual objects. For example, resizing, duplication, color or texture changing could be easily achieved for the benefit of efficient collaboration.

Although the contents of the working environment could be similar, the impression and perception in one's brain might vary dramatically.

In line with the physical presence factors mentioned before, this system provides full-scale image size. It could help immerse users with higher fidelity of the environment. On the other hand, each site is equipped with a tabletop with identical size. Thus, consistent image size and viewing distance could be maintained so that each user could perceive others the same way they are being perceived by others.

In addition to that, many of the interactivity factors are also supported in this groupware system. For example, the tabletop affords a number of characteristics that could be modified by the users. Both physical and digital manipulations are enabled via natural gesture instructions. Moreover, the tele-presence component further introduces channels that allow intentional information, such as facial expressions, gaze directions, and body movements, to be precisely perceived. These are all potential input sources for users to communicate efficiently and effectively.

Next, one of the key features, which this tele-presence component is designed to implement, is the ability to support high degree of correspondence between the type of user input and the type of medium response. Compared to common online chatting software with single camera (webcam) setup, this system enables full spatial faithfulness environment. Instead of sending identical video streaming to every remote user, as what is done by most conference software with single camera, this system sends different video streaming to others. Each copy is captured from the camera that embodies the eyes of the remote user. In that case, user inputs like gazes and movements can be properly represented with the spatial relationship precisely mapped. They will know who is getting attention, and who is paying, which cannot be easily achieved by single camera setup.

### 3.3 Tangible Augmented Reality

The significant feature of Tangible Augmented Reality (TAR) is to integrate the multi senses from haptic and visual stimulus. The combination of Tangible User Interface "TUI" and Augmented Reality (AR) attempts to enrich a user's real environment by adding spatially aligned virtual objects (3D models, 2D textures, etc) to it and manipulating the physical objects. This increases the sense of co-presence for people involving into the real involvement.

Firstly, from the visualization of prospective view, Augmented Reality technology (Anastassova, Burkhardt, Megard, & Ehanno, 2007) tried not to separate users from their physical environment by superimposing virtual digital content on their real world view since the Augmented Reality can merge the virtual and real worlds together seamlessly. In that way, it could eliminate the presence of display devices and maintain users their awareness of present environment. However, there are limitations for current technologies to make users completely eliminate the awareness of the existence technology. For example, the HMD does not provide the high resolution of the camera images. Therefore, the integrated systems of combining Tangible User Interfaces and Augmented Reality usually use high resolution of projections to display the images rather than the HMD. This benefits the large scale projection could offer a common reference for the users and enhance the sense of presence.

Secondly, TAR offers the advantages from TUI (Waldheim & Carpentieri, 2001) which opens another sensory of channels. TUI gives immediate feedback from physical interaction through the tactile clue. With the aid of TUI, the users can have natural interaction with their hands. It helps users to capture, recognize and understand human natural actions and signal such as hand gestures, eye gaze, body moment, etc. Particularly, it is essential to engage users' multi-sensory in the design activities. On the other hand, AR technology generates corresponding visual effects for those actions on the screen or in virtual world. The advantage of using AR is to enable virtual data appearing in the real physical world. TAR creates a environment which allows the users interact with virtual objects naturally by using normal tools as they were using with the real objects such as paddle, cup etc. The integration of TUI and AR technology does provide body language or other spatial cues, spatial presence and direct or peripheral awareness of the activity of participants. The direct manipulation with perceiving digital information can naturally arouse the human senses since the users always have the sense of "being there" which is related to fundamental concept of "presence". Figure illustrates the relationship for physical presence and social presence within the content of TAR technologies.

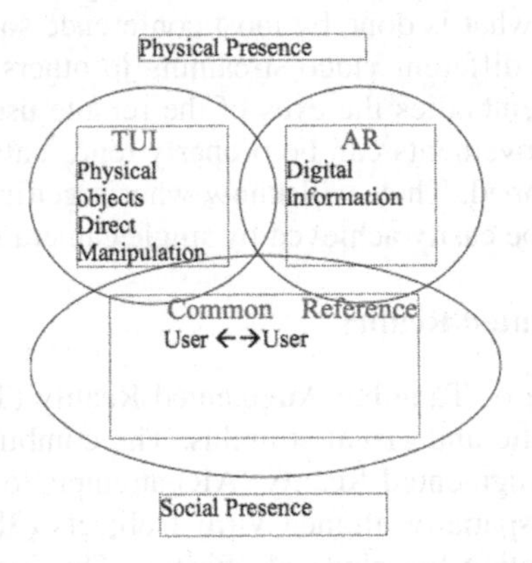


**Fig. 3.** An illustration of the relationship between physical presence and social presence within the content of TAR technologies.

For instance, the initial stage is that the individual user being the real environment physically and that brings out the direct feedback from visualization and tactile sensation. When the individual moves to the group work, they face to some direct interaction between each other with the level of social presence which also establish a sense of physical presence and increase of the sense of "being together".

# 4 Summary

As discussed in previous sections, the aim of this chapter is to introduce the concept and characteristics of co-presence and then use the concept to guide the design of Mixed Reality-supported collaborative systems. It is believed that the level of co-presence is an important factor that affects the design performance in collaborative design systems. After reviewing related literature, the key factors that could affect the level of co-presence have been investigated and discussed in the categories of physical presence and social presence. Although the influence those factors may have on the level of presence has not been measured in this chapter, they formed the basis of how Mixed-Reality-supported systems are conceptualized and specified. The significance of this chapter is that it has provided the concept and examples of specifying a mixed-reality supported collaborative system to increase the level of co-presence. The current prototypes of the designed systems are based on some design scenarios; however, the concept could also be applied to other collaborative systems and shared environments.

## References


Anastassova, M., Burkhardt, J.M., Megard, C., Ehanno, P.: Ergonomics of augmented reality for learning: A review. Travail Humain 70(2), 97–125 (2007)

Anderson, D.B., Casey, M.A.: The sound dimension. IEEE Spectrum 34(3), 46–51 (1997)

Anderson, J.D.: From jump cut to match action: An ecological approach for film editing. In: The Annual Conference of the University Film & Video Association, Philadelphia, PA (1993)

Barker, D.: Its been real - Forms of television representation. Critical Studies in Mass Communication 5(1), 42–56 (1988)

Bartle, R.: Early MUD History, vol. 2010 (1990)

Biocca, F., Levy, M.R.: Communication in the age of virtual reality. Lawrence Erlbaum Associates, Hillsdale (1995)

Biocca, F., Delaney, B.: Immersive virtual reality technology. In: Biocca, F., Levy, M.R. (eds.) Communication in the age of virtual reality, pp. 57–124. Lawrence Erlbaum Associates, Hillsdale (1995)

Dorr, A., Graves, S.B., Phelps, E.: Television literacy for young-children. Journal of Communication 30(3), 71–83 (1980)

Elliott, W.R., Rudd, L.R., Good, L.: Measuring perceived reality of television: Perceived plausibility, perceived superficiality, and the degree of personal utility, the Association for Education. In: Journalism and Mass Communication Annual Conference, Corvallis, Oregon (1983)

Everest, F.A.: Psychoacoustics. In: Ballou, G. (ed.) Handbook for Sound Engineers: The New Audio Cylopedia, pp. 23–40. Howard W Sams & Co, Indianapolis (1987)

Hatada, T., Sakata, II., Kusaka, H.: Psychoohysical Analysis of the Sensation of Reality Induced by a Visual Wide-Field Display. Smpte Journal 89(8), 560–569 (1980)

Heeter, C.: Being there: The subjective experience of presence. Presence 1(2), 262–271 (1992)

Held, R.M., Durlach, N.I.: Telepresence. Presence 1(1), 109–112 (1992)

Hellig, M.L.: El Cine del futuro: The cinema of the future. Presence 1(3), 279–294 (1992)

Ijsselsteijn, W.A., de Ridder, H., Freeman, J., Avons, S.E.: Presence: Concept, determinants and measurement. In: The Proceedings of Spie the International Society for Optical Engineering, vol. 3959, pp. 520–529 (2000)

IJsselsteijn, W.A., Freeman, J., Ridder, H.: Presence: Where are we? Cyberpsychology & Behavior 2001(4), 307–315 (2001)

Kim, T.: Effects of Presence on Memory and Persuasion. University of North Carolina, Chapel Hill (1996)

Kramer, G.: Sound and communication in virtual reality. In: Biocca, F., Levy, M.R. (eds.) Communication in the Age of Virtual Reality, pp. 259–276. Lawrence Erlbaum, Hillsdale (1995)

Lombard, M., Ditton, T.: At the heart of it all: The concept of presence. Journal of Computer-Mediated Communication 3(2) (1997)

Lombard, M., Reich, R.D., Grabe, M.E., Bracken, C.C., Ditton, T.B.: Presence and television - The role of screen size. Human Communication Research 26(1), 75–98 (2000)

Lombard, M., Reich, R.D., Grabe, M.E., Campanella, C.M., Ditton, T.B.: Big TVs, little TVs: The role of screen size in viewer responses to point-of-view movement. In: The Mass Communication Division at the Annual Conference of the International Communication Association, Albuquerque (1995)

Lombard, M., Ditton, T.: At the heart of it all: The concept of presence. Journal of Computer Mediated-Communication 3(2) (1997)

Milgram, P., Takemura, H., Utsumi, A., Kishino, F.: Augmented Reality: A class of displays on the reality-virtuality continuum. SPIE 2351, 282–292 (1994)

Neuman, W.R.: Beyond HDTV: Exploring subjective responses to very high definition television. In: Research Report for GTE Labs and the TVOT Consortium. MIT, Cambridge (1990)

Olds, A.: Small minds, high tech. Magazine of International Design 37, 54–57 (1990)

Potter, W.J.: Perceived reality in television effects research. Journal of Broadcasting & Electronic Media 32(1), 23–41 (1988)

Reeves, B., Detenber, B., Steuer, J.: New televisions: The effects of big pictures and big sound on viewer responses to the screen. In: The Information Systems Division of the International Communication Association, Washington, D.C (1993)

Sallnäs, E.-L. (ed.): Collaboration in Multimodal Virtual Worlds: Comparing Touch, Text, Voice and Video. Springer, New York (2002)

Short, J.W.E., Christie, B.: The Social Psychology of Telecommunications. Wiley, London (1976)

Steuer, J.: Defining virtual reality: Dimensions determining telepresence. In: Biocca, F., Levy, M.R. (eds.) Communication in the Age of Virtual Reality, pp. 33–56. Lawrence Erlbaum Associates, Hillsdale (1995)

Waldheim, L., Carpentieri, E.: Update on the Progress of the Brazilian Wood BIG-GT demonstration project (2001)

Wang, R., Wang, X.: Mixed reality-mediated collaborative design system: Concept, prototype, and experimentation. In: Luo, Y. (ed.) CDVE 2008. LNCS, vol. 5220, pp. 117–124. Springer, Heidelberg (2008)

# Part III
# Design Cooperation: Sharing Context in Collaborative Virtual Environments

Collaborative Design in Virtual Environments at Conceptual Stage
Walid Tizani (University of Nottingham)

"Scales" Affecting Design Communication in Collaborative Virtual Environments
Jeff WT Kan, Jerry J-H Tsai and Xiangyu Wang
(Taylor's University College, Yuan Ze University & University of New South Wales)

Design Coordination and Progress Monitoring during the Construction Phase
Feniosky Peña-Mora, Mani Golparvar-Fard, Zeeshan Aziz and Seungjun Roh
(University of Illinois at Urbana-Champaign)

# Part III
# Design Cooperation: Sharing Context in Collaborative Virtual Environments

# Collaborative Design in Virtual Environments at Conceptual Stage

Walid Tizani

The University of Nottingham, UK

**Abstract.** The conceptual design stage of a construction project has significant consequences on all the stages that follow. This stage involves multi-disciplinary design activities that would benefit from dedicated support from information technology systems. The use of real-time multi-disciplinary collaborative systems supporting design teams is still the subject of much research. The most challenging requirements for such systems are the design of a 'collaborative-aware' information models, the implementation of concurrency in design, and the management of the design processes and workflow necessary for multi-disciplinary design teams. This chapter outlines the requirements of collaborative systems and briefly describes an experimental collaborative design environmental. It also proposes methodologies for the issues of concurrency and the management of processes. Concurrency of design was done through the automation of the synchronization of a shared information model. The management of design processes was done through controlling access-rights of designers to the shared information model.

**Keywords:** design, collaboration, multi-disciplinary, concurrency, access-rights.

## 1 Introduction

The design phase is primarily concerned with specifying the 'product' that best fulfils the client's brief, ensures safety during construction and use, and achieves minimum overall cost. This process requires interactions, primarily, between the disciplines of architecture, building services, structural engineering and site construction. Each specialist within these disciplines has specific understanding of the design problem. This compartmentalized decision-making can cause problems with downstream design activities, and conflict with upstream decisions and intentions. So, although the design process itself constitutes around 5% of the costs associated with a typical construction project, its decisions impact on the build cost and the quality of the remaining 95% (Egan, 1998). Therefore, the design process, and more specifically the conceptual design process, is a critical part of any project and supporting it is an important factor in improving the overall life cycle of construction projects.

The fast advances in information technology have greatly assisted collaboration at the design stage by mainly facilitating communication between the various

X. Wang & J.J.-H. Tsai (Eds.): Collaborative Design in Virtual Environments, ISCA 48, pp. 67–76.
springerlink.com 

designers. This is done through document management systems using web technologies or interoperability of software through the use of neural file formats. The much sought goal of real-time collaborative design systems is still largely experimental. This is due to the need to overcome not only technological issues, such as effecting concurrency, but also the complexity of the multi-disciplinary design processes and workflow.

This chapter attempts to shed some light into the technological issues related to the implementation of collaborative systems through the description of the main challenges that have been tackled when designing and implementing an experimental real-time and multi-disciplinary collaborative system.

## 2 Importance and Complexity of Conceptual Design

Addressing design requirements form the point of view of a single discipline is a major creative task given the open-ended nature of it. The design for a construction facility is further complicated by the multi-disciplinary nature of it and the fact that each discipline adds its own sub-set of requirements and constraints. The multi-disciplinary design process is thus a fluent and complex process that attempts to satisfy discipline-based, cross-discipline-based as well global constraints and requirements. A successful design process should address the multi-disciplinarily nature of the problem so as to minimize cross-discipline conflicts while at the same time not losing sight of the original overall requirements.

Much communication between the disciplines is normally necessary in such a process. The communication deals with the specification for the various parts of the overall design with much iteration. The process carries its own momentum in that each specification set is built upon by others in such as way that it is difficult to roll back the process and start again and each different initial start might results in different final design. It is not unknown that a final design might not adequately fulfil the original intentions. One of the main causes of such variation from the original intentions is the inability to easily assess the consequences of design decisions of one discipline onto the others. The specification provided by one specialism tends to focus primarily on its own imposed constraints. This is not due to unwillingness to cross-check with other affected specialisms but often due to the complexity of doing so.

The complexity of the multi-disciplinary design process and the importance of such a process in its effects on the quality and cost of a project make supporting the collaborative design process a key for improving the design outcome. This has also become more vital due to the increasing prevalence of distributing the design activities and the necessity to work remotely.

## 3 Requirements of Collaboration Design Environments

The aim of a virtual collaborative design system is to provide designers with a set of mechanisms and tools for specifying and building-on shared information in a coordinated and organized manner with the goal of agreeing on a solution that meets a set of constraints. Central to the design of such a collaborative environment is extracting the system high-level requirements based on the needs of the collaborative and concurrent design process. The following set is considered to include such requirements (Fahdah and Tizani, 2008):

- *Concurrency*: Concurrent design is a key aspect of real-time collaborative design. The updates to a shared model need to be managed so to allow for such concurrent design activities.
- *Information Modeling*: A product model in a shared workspace needs not only to cover the data representation of the building but it also needs to support collaboration processes and workflow for multi-disciplinary applications where the data from different disciplines are greatly interrelated.
- *Access Control*: Collaborative design environments, in general, and integrated design environments, in particular, would require careful management of access-rights to modifying shared data models. This is a key measure for managing multi-disciplinary design processes.
- *Version Management*: In concurrent collaborative design systems where there is one version of the model shared by all designers, version management is required to support the restoration of the design data to a previous stage in cases such as dead-end design or design conflicts. Version management also needs to allow for tracking of the changes made to the product model elements for auditing purposes.
- *Communication Tools*: Collaborative design systems should be supported by communication tools. Examples of these tools include Email, notification facility, discussion boards, shared whiteboard, instance message exchange, visual graphical discussion board, and video-conferencing.
- *Intuitive Interface*: 3D virtual models allow designers to intuitively access and modify the product model data. Such an interface should include the generation of virtual 3D graphical representation of the product being designed and allow for selective viewing of part of it to reduce complexity and to aid the design process.
- *Performance*: The data transferred in a collaborative design system is expected to be huge and complex. The performance of the system is an important factor of a real-time collaborative design system so to avoid networking becoming the bottleneck of the design process.
- *Design Automation*: The provision of sufficient design task automation would save the engineers conducting low-level designs and reduce human error. The automation requirement in a collaborative design system is more crucial for decision-making than collaboration.
- *Document Management*: The number of documents in a typical project would be large. It is, thus, essential to have a facility to manage such documents.

The most challenging of the above requirements are the first three. The following sections describe experimental methods and technologies that have been used to address two of these challenges namely: concurrency and access control. This is tested through the development of an experimental collaborative environment for the design of multi-storey steel structures.

# 4 An Experimental Collaborative Design Environment

This section describes an experimental collaborative environment for the design of steel structures (Fahdah and Tizani, 2006; Fahdah, 2008). The environment is a client-server application composed of a central server, maintaining the shared design model, and multi-client workstations accessing the shared model via network.

The prototype application has been developed to encapsulate the design approaches using 'software agents'. All processes and product models are implemented in terms of objects logically interconnected, using object-oriented methodology and .Net technology. The agents in the systems are implemented as active objects which encapsulate their own thread of control, identity, state, and behaviour. The communication infrastructure was implemented using ".NET Remoting". .NET Remoting enables different applications to communicate with one another, whether those applications reside on the same computer, or on different computers. The environment operates in three modes: standalone, over an intranet, or over the Internet. When operating over a network the client agents communicate with the server using TCP/IP protocol if it is a local network and HTTP protocol if it is the Internet.

## 4.1 The Virtual Model

The environment has a "virtual prototyping" interface which was developed in C++ and OpenGL (Open Graphic Library) to create a real time dynamic system. Default views are provided for each of the traditional roles including the architect, structural designer and services designer. Model manipulation is specialised for each of the views so that no information overload will occur. However, each view can be customized to visualise any of the options available to others by overriding the default set.

The graphical representation of design data is achieved through linking the graphical methods to the product model. Designers are therefore able to manipulate the product model through interacting with the graphical model. This can assist in bridging the gap between the various discipline-based representations in an intuitive fashion.

## 4.2 Description of a Typical Design Cycle

The environment allows for the incremental development of a shared Information Model starting from any design or discipline stage. However the description of its capabilities is better first described using a typical sequential design process. Figure 1 gives an overview of the main stages of a possible design cycle (steps 1 to 9). The sequence shows:

1. The project manager creates the project space (and with it a shared model on the server) and defines the design team members (actors) and their roles (client, architect, etc.). The actors are granted default permissions based on their respective roles. Roles, Actors and Permissions are explained in Section 6.

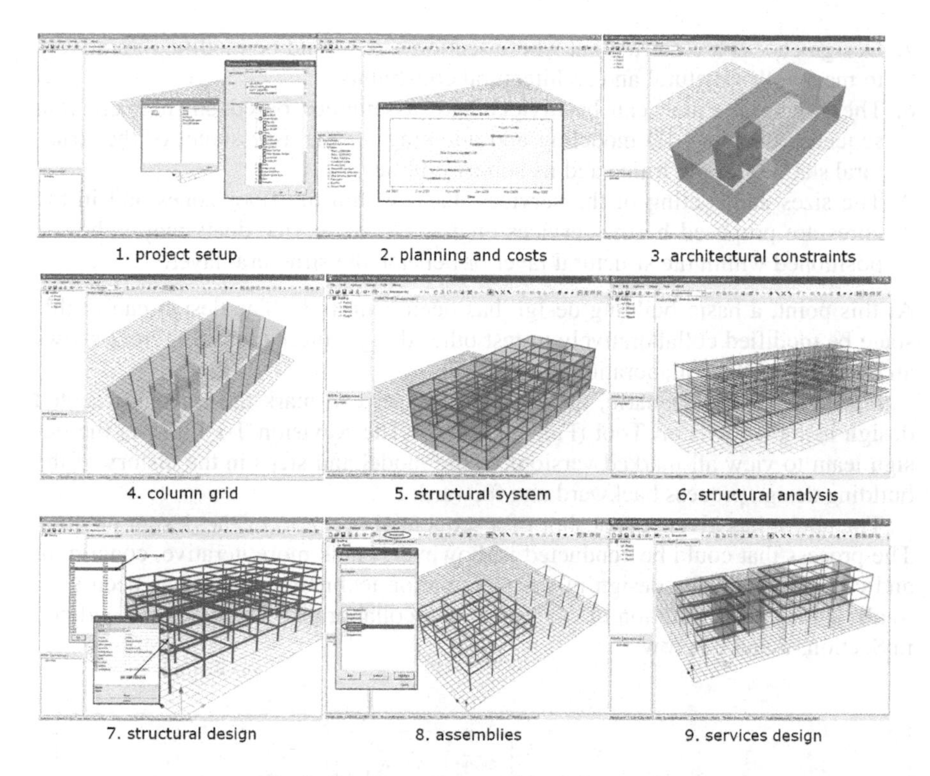


**Fig. 1.** Sequence of a possible design cycle.

2. The client defines the building's overall requirements. This includes the building's targets such as costs for different parts of the project, total floor space, and completion dates for the main activities.
3. The architect specifies the layout of the building's perimeter, cladding system, floor area designations, the number of floors, and the internal height requirements for each floor components (structural and others). The architect may also impose column positioning constraints.
4. The architect or the structural engineer proposes a grid spacing to meet the imposed constraints.
5. The structural engineer adds the flooring system, initially using automated sizing algorithms that apply the loading implied from the area designation imposed in step 3. The structural frame is complete with the addition of the lateral bracing system. At this stage, a complete structural prototype that meets the architectural constraints would have been obtained.
6. The prototype now has sufficient data to produce a 3D structural analysis model of the building. This is generated by a *Structural Analysis Agent* to accurately model the members and connections with the appropriate use of analysis members and dummy members to model connection offsets, hence maintaining compatibility between the as-built structure and the analysis model. The analysis is carried out to produce the structural analysis response model.

7. Design checks are carried out where section sizes could be modified iteratively to meet both structural and architectural constraints.
8. The structural frame can be divided into assemblies based on their erection sequence enabling 4D modeling and allowing for the assessment of the structural stability of the frame and its sub-assemblies.
9. The sizes and routing of the services ducts within the floor zones and in the cores are proposed by the services engineer. The service ducts may either be positioned within the structural layer or beneath the structural layer.

At this point, a basic building design has been achieved. The design can at this stage be modified collaboratively to test other design alternatives. Figure 2 shows an overview of the collaboration model.

For any redesign scenario, the project manager can mark the current accepted design using a Revision Tool (Fahdah, 2008). The Revision Tool allows the design team to view all marked versions of the model and steps in the history of the building design process backward and forward.

The above has given a snapshot of a sequential scenario of the design process. The process that could be conducted is, however, much more iterative, concurrent and collaborative. The design of the information technologies that has been used to allow for implementation of concurrent and collaborative processes are outlined in Sections 5 and 6 below.

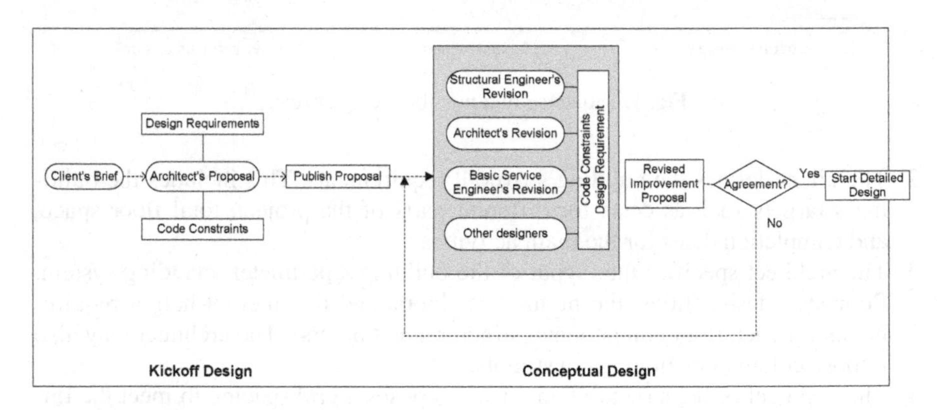


**Fig. 2.** Overview of the collaborative model for conceptual deisgn

# 5 Communication Approach and Concurrency

In a real-time collaborative environment, concurrent access to shared resources is a key feature to supporting collaboration. However, providing concurrency in building design is a complex task as that may lead to conflicts and data inconsistency.

Two main different strategies can be adopted for the update of shared data models: synchronous or asynchronous (Li and Qiu, 2006). In an asynchronous paradigm, the design activities are managed locally and then coordinated at assembly level. This is normally only suitable for applications where the design

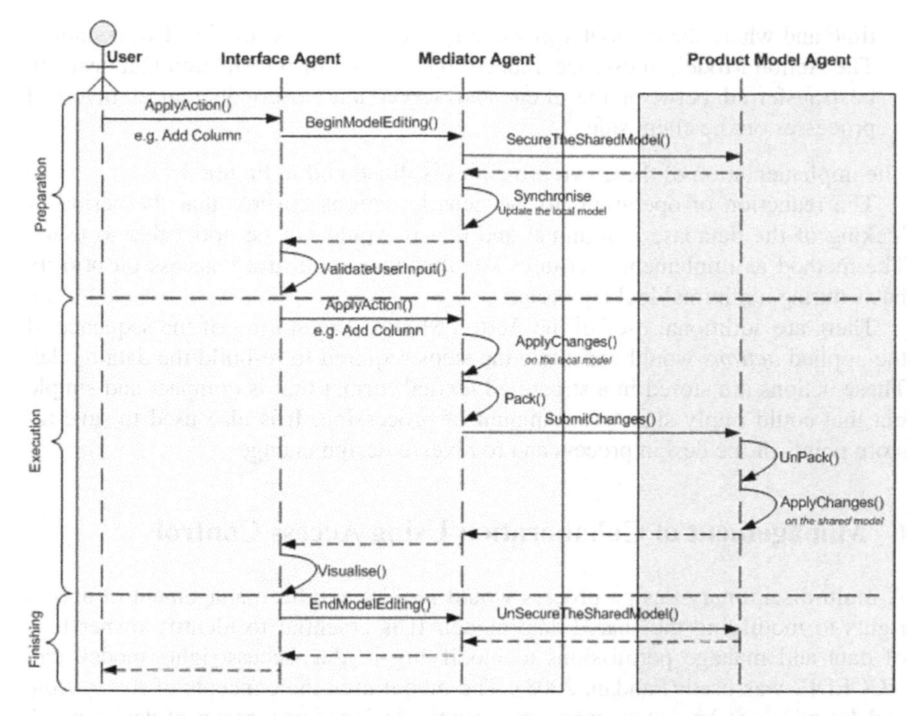


**Fig. 3.** Communication diagram showing the execution of one action.

tasks are not highly interrelated. In the synchronous paradigm, the design activities can be managed concurrently and is more suitable for collaborative design. However, this method requires a suitable data transaction management facility and data merging techniques that ensure data consistency and avoid conflicts.

A pessimistic approach to the synchronous paradigm is often used to maintain integrity. This is done by allowing only one the write access to the data at a time. That is the central data model is locked for changes but for the current designer (Paulraj, 2003). This may lead to delay in achieving tasks especially if data is left exclusively locked for too long.

For the experimental environment described here, a Data Transaction Management Model has been designed to alleviate the inefficiencies of the pessimistic method and to affect a concurrent design process. The adopted model is based on two principles:

1. *Automating the synchronisation method to the shared data model.* This is done by providing automatic methods to ensure that only a single designer can update the shared model at any given time. The shared model database is locked prior to modifying any data stored in the database, and unlocked once the modification is complete.
2. *Minimizing the locking period of the shared model.* This is done by implementing the so-called "Action Model" where any computing operation is reduced to its generic actions (Fahdah, 2008). The generic actions are acted upon one at a

time and where the computer processing is carried out at the local workstation. The Action Model, in essence, reduces the amount of information that needs to be transferred between the client and server and to conduct most involved processes on the client side.

The implementation of these two principles is illustrated in Figure 3.

The reduction of operations to the generic actions ensures that the period of locking of the database is minimal and that it would not be noticeable to users. The method as implemented ensures simultaneous multi-users access on priority basis during a minimal locking period.

There are additional uses of the Action Model. The storing of the sequence of the applied *actions* would constitute the steps required to re-build the data model. These actions are stored in a structured textual format that is compact and simple but that could imply significant amount of processing. It is also used to save re-store points of the design process and to reverse design changes.

# 6 Management of Collaboration Using Access Control

A multi-disciplinary design process would require careful management of access rights to modifying the shared data model. It is essential to identify ownerships of data and manage permissions to modifying it. An Access-rights model, call ACCEDE, was used (Fahdah, 2008). The model uses the concepts of *Actor*, *Role* and *Permission*. An *Actor* represents a single designer or a group of designers. A *Role* could be seen as representing a discipline with what this implies from scope and responsibility within a project. The default roles are the traditional roles within the construction industry, i.e. Client, architect, structural engineering, etc. The *permission* concept is the key control mechanism regulating access-rights of actors to any part of the data model. Permissions in the system can be seen as the link between the engineering data representation and the collaboration management system through actors and roles. The generic types of permissions are r*eal-only*, *delete*, *create*, *ownership-change*. Figure 4 shows the schematic view of the relationships between the model elements, the actors, and the permissions.

These concepts of *Actor*, *Role* and *Permission* are integrated with the product model to allow access controls or *Permissions* to be applied at different levels: *type-level* (e.g. columns), *element-level* (e.g. a specific design element) and *attribute-level* (a specific attribute of a type of element).

The above implementation will regulate the multi-disciplinary and inter-disciplinary collaborative design processes. Due to the complexity of managing the allocation of access-rights, a default implementation exists for new projects complete with roles, actors and type-level permissions. More specific permissions such as element-level and attribute-level is under the control of the owner of the data or resource where the default owner of an element is its creator. Owners can adjust the default general permissions to allow other actors to access their resources. The system propagates the currently applied default permissions to new design elements and for new actors.

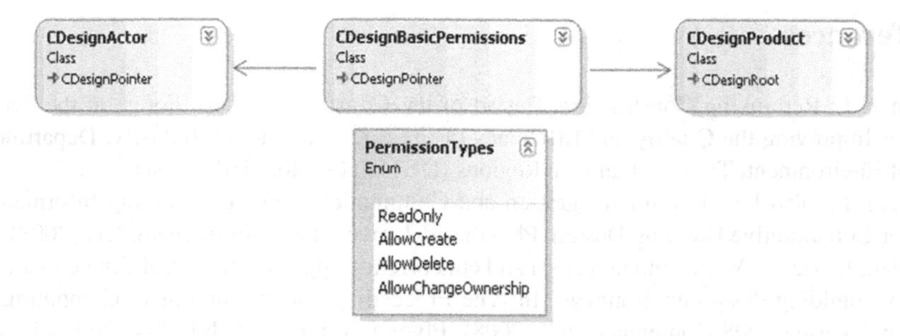


**Fig. 4.** Schematic representation of the link between the Product model and access rights

The *Role* concept is used to allow the system to grant default permissions to the designer(s) to access the product model elements based on their assigned roles. For example, project managers have full access rights, clients have read-only rights, architects and engineers will have access rights based on the design tasks distribution that is initially based on their traditional division of responsibilities. However, granted permissions can be altered afterward.

The above, briefly outlined, the design of an access-rights model has been found successful and sufficient for the management of the collaborative design process (Fahdah, 2008). It differs from other implementations by providing users with more control of their own input.

## 7 Summary and Conclusions

The conceptual design stage of a construction project has significant consequences on all the stages that follow. This stage involves multi-disciplinary input that would benefit from dedicated support from information technology systems. Work in this area has led to advances in the use project document-management systems and in the use of inter-operable software. The use of real-time collaborative systems supporting multi-disciplinary design team is still the subject of much research.

The most challenging requirements for such real-time collaborative systems were outlined as the design of a 'collaborative-aware' information model, the implementation of concurrency in design, and the management of the design processes and workflow necessary for multi-disciplinary design teams.

An experimental collaborative design environment was briefly outlined. Through the design and implementation of this environment, proposals were made on possible models for the issue of concurrencies and the management of collaboration using an access right model.

The experimental software has provided insight into possible solutions for the main challenges posed by the implementation of multi-disciplinary collaborative environments. There remains that such information technologies can be proven in the applied field and be generalised to apply to wider application areas.

## References

Egan, S.J.: Rethinking Construction, Report of the Construction Task Force on the Scope for Improving the Quality and Efficiency of the UK Construction Industry, Department of Environment, Transport and the Regions (DETR), London, UK (1998)

Fahdah, I.: Distributed IT for Integration and Communication of Engineering Information for Collaborative Building Design, PhD thesis, University of Nottingham, UK (2008)

Fahdah, I., Tizani, W.: Communication and concurrency approach for a real-time collaborative building design environment. In: The Proceedings of the Intelligent Computing in Engineering 2008 Conference (ICE 2008), Plymouth, UK, CD, July 2-4 (2008), ISBN: 978-1-84102-191-1

Fahdah, I., Tizani, W.: Virtual collaborative building design environment using software agents. In: ISSA R. (ed.) Proceedings of the 6th International Conference on Construction Applications of Virtual Reality, Orlando, Florida, USA (2006)

Li, W., Qiu, Z.M.: State-of-the-art technologies and methodologies for collaborative product development systems. International Journal of Production Research 44(13), 2525–2559 (2006)

Paulraj, P.: Database Design and Development: An Essential Guide for IT Professionals. John Wiley & Sons, Chichester (2003)

# "Scales" Affecting Design Communication in Collaborative Virtual Environments

Jeff W.T. Kan[1], Jerry J.-H. Tsai[2], and Xiangyu Wang[3]

[1] Taylor's University College, Malaysia
[2] Yuan Ze University, Taiwan
[3] The University of New South Wales, Australia

**Abstract.** This chapter explores the impacts of large and small scales of designed objects towards the communication in three-dimensional collaborative virtual environments. The motivations are twofold: 1) to better understand design communications in virtual environments; 2) to suggest improvements of present virtual environments so as to better support design communications.

**Keywords:** design communication, protocol analysis, scale, virtual environments.

## 1 Is Scale an Issue in Collaborative Virtual Environments?

In many cases, the primary motivation of people using the virtual environments is the desire to socially interact with a different identity. With the advance of technology, social virtual environments are becoming popular. Nowadays virtual environments have a good sense of immersion, hence the sense of presence (Shanchez-Vives and Slater, 2005). Linked with the ability of multi-user modelling, these environments provide a good medium for distant design collaboration.

Consider the following scenario, which is adapted from the communication of an empirical experiment in a 3D virtual environment, between two designers – Jack and Iris:

Jack: You see... where are you? [looking around]
Iris: I'm up there, I can see you [fly near to Jack]
Jack: You're flying!
Iris: To get a better view of what we're doing...
Jack: ... never mind, you're not seeing what I'm seeing...

Several issues occur in this communication. Firstly, because of the scale of the project, Iris needs to "fly up" to get an overview or a bird's eye view of the project. Essentially, she was using the vertical distance to scale down the "object" they were asked to design. Secondly, in design collaboration, one important aspect of communication is to get cognition synchronized – knowing and understanding what the others are thinking. In order to achieve this, they need to share the same context during communication (verbal or non-verbal). The last utterance of the

X. Wang & J.J.-H. Tsai (Eds.): Collaborative Design in Virtual Environments, ISCA 48, pp. 77–87.
springerlink.com 

above example exemplifies how the communication was affected by not seeing the same scene. This was caused by Iris leaving without notifying Jack, in turn this was caused by the scale of the design.

When designing, architects scale down the buildings to get the topological layout, while electrical engineers scale up to circuits to fit in the details. Using different scale to design comes very natural to designers. When CAD was introduced, due to the limited display areas especially in the early years, zooming and panning became two of the mostly used commands. These two commands closely related to scale.

Normal people know how to read the environment (large-scale) and how to handle objects (small-scale). These two abilities seem to require different cognitive resources. We observe the environment by sensors. Visually, we move our eyes, head or even the body. We also sense the environment by touch, smell and listen. Small objects provide the luxury for us to hold, rotate, and experiment with it. We have learned to handle small objects and transform it into tools. We seem to use different resources to decode these small objects that can be handled and manipulated by our hands. This behavior is even observed in infants. When they are introduced into a new environment (large-scale), they move to observe the environment. However, when they receive new toys (small-scale), they experiment it with their hands – bang it, try eating it, etc.

In virtual environments, software tends to borrow metaphor from real environments to navigate, such as walk, fly and drive. Interestingly, most software has an object mode to explore small-scale objects, for example in the 90's the Quicktime VR (Virtual Reality) handled environments (large-scale) and objects (small-scale) very differently. The former uses 360 degree photos stitched together while the latter flips a sequence of photos of an object taken at different angles.

In this chapter, we explore the impact of scale upon design communication in real and virtual environments and suggest some possible improvements. This chapter commences from background studies including design communication and protocol analysis. It is then followed by the illustration of our previous studies. We then focus on the discussion of the impacts of different scales on design communications in Virtual Environments.

## 2 Background

This section introduces the background of our investigation method. We use the conversation, communication, among the collaborators as a non-invasive *in-vitro* mean of capturing data.

### 2.1 Design Communication

Whenever there is a design task that involves more than one party, communication is unavoidable. Research of communication has been conducted in various disciplines such as linguistics, social psychology, and information theory. Bull (2002) argues that communication can be studied in its own right. Management studies (Allen et al., 1980; Tushman and Katz, 1980) show that project performance in a variety of organizational settings is positively related to communication and information exchange. However, in the study of communication and decision-making, Hewes (1986), based on socio-egocentric theory, claimed that the content of social interaction in small groups does not affect group outcomes, rather those non-interactive *inputs* factors are more important. The *inputs* are:

> "*shared variables that individual group members bring to a discussion, such as cognitive abilities and limitations, knowledge of the problem or how to solve problems individually or in a group, personality characteristics, motivations both individual and shared, economic resources, and power*" (Hewes, 1996 p181).

Notwithstanding Hewes's claim, many others such as Minneman (1991) believe social process is an important part of the group process. Minneman (1991) argued design work emerges from the interactions of the group to establish, maintain, and develop a share understanding. He suggested that designs are created through an interactive social process. Stempfle and Badke-Schaub (2002) recorded and analyzed three teams' communications and found that teams spent about 2/3 of their interaction on the content and 1/3 on the group process; also teams spent about 10% of their content-directed activity on the goal space, whereas remaining 90% was focused on the solution space. Their study, and other studies on design collaborations (Cross and Cross, 1996; Olson and Olson, 2000; Oslon et al., 1992; Sonnenwald, 1996; Zolin et al., 2004), demonstrated the importance of communication and social processes of collaboration.

Design communications in collaborative design can be viewed as a mechanism for designers to clarify the design goal, to understand each other's design ideas, and to propose design strategies. With the advance of information technology, collaborative 3D virtual environments are available to allow synchronized collaboration across different geographical locations. Virtual environments can provide channels that bridge some limitations of real environments, however, they also post other limitations. Currently, the majority of studies of design collaboration in virtual environments mainly focus on analyzing collaborative design behaviors in virtual worlds (Maher et al., 2006; Gul et al., 2008; Gabriel and Maher, 2002). The studies of collaborative design behaviors affected by different scales of design projects are rare. We used protocol analysis, as a method, to study the impact of scale upon design behaviors.

## 2.2 Protocol Analysis

Ericsson and Simon (1993) argued the recording of talking aloud or concurrent reporting – verbal protocols – can be treated as quantitative data for studying thought process. Van Someren et al. (1994) provided a theory background and some practical guide to study and model cognitive process. They assumed a simple model of the human cognitive system, as depicted in Figure 1, to develop the validly of concurrent reporting. The arrows in the diagram represent five different processes: perception (sensory to working memory), retrieval (long-term memory to working memory), construction (within working memory), storage (working memory to long-term memory), and verbalization (working memory to protocols) (Van Someren et al., 1994).

Before analysing the protocols, it is important to know what cognitive processes are to be studies because it is impossible to model the entire cognitive processes. In general there are three major procedures in protocol analysis: data collection (audio, video, and artifacts), data organization (transcribing, segmenting, and encoding), and data interpretation. Tang (2002) classified the procedures into five steps: conducting experiments, transcribing protocols, parsing segments, encoding according to a coding scheme, and interpreting the encoded protocols.

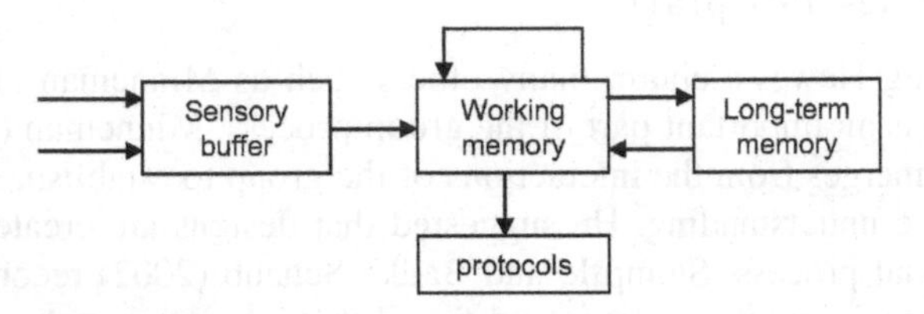


**Fig. 1.** Memory model for the validly of protocol analysis (Van Someren et al., 1994).

Before all these theoretical development, in the domain of design, Eastman (1970) conducted the first formal protocol analysis that studies the process of design. He viewed designing as a process of identifying the problems and testing alternative solutions.

In a collaborative design environment, it is impossible for individual member to think-aloud, however, Cross et al. (1996) suggested that

> *"The verbal exchanges of members of a team engaged in a joint task seem to provide data indicative of the cognitive activities that are being undertaken by the team member."*

Many researchers (Goldschmidt, 1995; Gabriel and Maher, 2002; Maher et al., 2006; Gul et al., 2008; Tsai et al., 2008) took this view and used protocol analysis technique to study design teams. They mainly treated their verbal communication as a nature form of talking aloud and considered them as the raw protocol data.

Protocol analysis has limitations in capturing the non-verbal thought processes going on in the design process, especially in a teamwork environment. Therefore, important non-verbal communication is often neglected. However, Cross (1996) suggested it is a viable, if not the only method available, to scientifically study designers.

## 3 Study of the Impacts of Different Scale on Design Collaboration in Virtual Design Studio

Tsai et al. (2008) conducted a study of different scales of collaborative design in real and virtual environments. Four design projects were implemented for the study that covers design collaborations with four variables: 1) large-scale design projects, 2) small-scale design projects, 3) real environments, and 4) virtual environments. The large-scale design is defined as a building, which people can move within it (see Figure 2). It consists of different compositions of spaces, for example, different spatial arrangements of public spaces and private spaces. In contrast, the small-scale design is defined as a design of furniture which people can use and move it but cannot move within it, as shown in Figure 3. This behaviour is like the objects that can be handled by hand discussed in Section I. A studio and a workstation are designed in the large-scale design project and the small-scale design project respectively. Both of them are performed in the real environments by face-to-face design collaborations and also in the virtual environments in Second Life (http://secondlife.com/). A coding scheme categorized into communication control, design communication, social communication and communication technology for protocol analyses was developed to analyse and compare design implementations among the four design tasks. In general, design communications has the highest occurred percentage compared with communication control, social communication and communication technology in all four design projects.

Tsai et al. (2009) further focused the study of different design scales on design collaboration in 3D virtual environments with one studio design project (large-scale) and one workstation design project (small-scale) in Second Life. Figure 2 shows the screen-shots of the large-scale and small-scale projects. Some main findings from protocol analyses of different scales impacting on design collaboration in 3D virtual environments are as follows:

In both large-scale and small-scale design projects:

- There is higher percentage of "Pause" in communication control compared with "interruption" and "hand-over". Pause represents a temporary cessation of conversation between the group members. Interruption is for a participant interrupts another participant and is associated with simultaneous speech (Levinson, 1983). Hand-over is to hand over the conversation of a speaker to another participant. One of the main reasons is that design group members spend a lot of time on modelling objects in Second Life.
- In design communication, design task has the highest percentage compared with design concept and design detail. In general, design group members are concerned about how the design is implemented. Design Task is about what actions are taken to get the task done in accordance with the brief, e.g., task questing, instructing and working status. Design concept is about how design ideas are manipulated during the design process. Design detail is how the component of the design project is created and modified in virtual environments, e.g., size, shape, colour and texture of 3D objects.
- In 3D virtual environments, in Second Life, design group members can easily change colour and texture of objects and see the results instantly. "Colour and texture" is discussed mostly in design detail.

(a)

(b)

**Fig. 2.** (a) large-scale design, multi-level studio; (b) small-scale design, workstation.

However, in the large-scale design project, which is a studio design:

- Design group members spend time to explore the environment and space according to their own preferences. There are fewer interactions between design group members.
- Design concept is concerned more than design details.
- Task questioning is highly concerned in design task. It is due to the large-scale design area is much border than the small-scale design. Design group members are dispersed and they need to coordinate and manage design tasks with each other.
- Design group members are scattered at different levels, therefore, they are not able to see the avatars of each other. Therefore, there is no point to gesture.

In the small-scale design project, which is a workstation design:

- Design group members are forced to focus on the object. There are more interactions between design group members.
- Design group members have higher percentage of "working status" communications for design task. This indicates design group members are more interactive in their collaboration because they informed each other who concern the tasks.
- Design project members can see each other's avatars most of the time. Therefore, they can use the gesture function in Second Life as a mean for social communication.

## 4 The Impacts of Different Scales on Design Communications in Virtual Environments

The coding schemes of the above two studies (Tsai et al., 2008; Tsai et al., 2009) covered communication in a broad range. In this section, we focused the discussion in design communication. We define design communication as:

1) any communications that is related to the objects being designed; and
2) any communications regarding to activities associated with the first clause.

Within this design communication boundary, there are two basic types: verbal and non-verbal. Only the first clause of the definition has direct impact on the design artifact. The second clause plays a supportive role. People can spend all the time planning, scheduling, dedicating but not doing the design work. On the other hand, people need to know who is doing what and also when it should be done. Section 3 touches some of the impact of scale upon verbal design communications. In the following discussions, we focus on the impacts of different scales on the design communication content and non-verbal design communication in virtual environments.

## 4.1 Impact of Scale on Content of Design Communication

In the large-scale project, from the result of Tsai et al (2009), participants' design communications were mostly concerned about who is doing what (task questioning). It occupied nearly 30% of the total design communication. The small-scale had only about 17% of task questioning. This belongs to our second category of design communication – design supportive communication which indirectly influences the design product.

We conjecture, with everyday experiences, in a collaboration we need to know where the others are and their contributions to be productive. Friedman et al. (2007) also indicated the significance of proxemics towards behavior of avatar in Second Life.

In a large-scale project, it will be natural that collaborators disperse to their "own area" which they need to design. Socially, this may or may not be desirable. It will likely to be undesirable if there is a lack of trust. It will be desirable if the project can be decomposed into smaller pieces that everyone in the team can work in parallel. We further conjecture for new teams to work efficiently with a large-scale project in a virtual environment, they need to engage a lot in the second category of design communication, that is not directly related to the design objects. The success of the project will depend upon the support of this kind of communication, such as proximity, work progress and status.

In a small-scale project since the team will be gather around the design objects, the issue of proximity will not be a big concern. The issue becomes how to effectively support the first category design communication – directly related to the design object.

## 4.2 Impact of Scale on Non-verbal Design Communication

There are two basic different types of non-verbal communication in the physical world: storable and non-storable. The storable communication includes sketching, writing, and model making. The non-storable communication includes gestures, which can be stored in a virtual environment. The impact of "scale" on those storable non-verbal communicate are obvious – they scale it down and make sure everyone is working on the same scale.

On the non-storable side, when collaborating face-to-face to design, people gesture differently when the size of the thing being designed is different. In the real environment (one-to-one scaled environment) if the designed objects are within the reach of our body, the movement of people's gestures usually reflect the real size of the thing being designed. However, if the "scale" is too big, people will use finger pointing, usually in conjunction with a scale downed plan. Even on the drawing board, when the designed objects are being scaled down, people will use real size gesturing when the objects are small, for example the location and size of a window in proximity to the users.

In studying design collaboration, one of the early paper by Tang (1991) observed that a high percentages (one third) of the "drawing space activities" were gestures. Visser (2007) developed a description language for graphico-gestural to study design collaboration in terms of representational and organizational activities. In our discernment, organization gestures are not strictly design related.

We define design-related gestures into three types. By observing tapes of design collaboration gestures for design communication are classified into the following types:

- functional: related to meaning,
- behavioral: related to behavior of environment, user or designed object, and
- structural: related to shape, form or location.

The behavioral gestures usually express movements, for example people, sun, wind, etc. The structural gestures usually represent the form of the designed object. The functional type gestures are comparatively rare and may be used in conjunction with other types of gestures, for example, a designer uses his fist to hit the drawing to signify a place of importance. It is both functional and structural because it conveyed the both location and the meaning of the location.

There is one interesting observation that gestures do not only serve the purpose of communication, but it seems to be an integral part of designing. We observed blindfolded designers gesturing when they were designing. We also observed designers using structural body gestures in computer-mediated collaboration even when the other participants cannot see their gestures.

### 4.3 Suggested Guidelines for Further Improvements in Virtual Environments to Support Design Communication

Based on the studies of the impacts of different scales design projects on real and virtual environments as well as non-verbal design communications, the following proposes suggested guidelines for further improvements to support design communication in virtual environments:

- Scaling: Tsai et al. (2009) reported that when designing within a "large-scale" project, participants were scattered and the collaboration was mostly loosely coupled as defined by Kvan (2000). On the other case, in the "small-scale" project, participants could be closely coupled and work together. In face-to-face design collaboration, scaled models, drawings, and computer screens were used as tools to facilitate closely coupled design communication. We propose to implement tools to allow users to work with different scales within the environment so that collaborators can look at the "large-scale" object in a "small-scale" manner.
- Gesturing: The gesturing capability of current virtual environments supports mainly social gesturing of avatars. In the study by Tsai et al. (2009), there was no gesturing in the large-scale project and the gesturing in the small-scale project were all social non-design related. We suggest future environments need to

implement at least spatial/structural gesturing for users to communicate the sense of location, size, and shape.

- Presence: Task questioning occupied a third of the design communication in the study by Tsai et al. (2009). We hypothesize that if they know where the others are and what they are doing, this type of communication will be lessened and they could focus on the design objects.

## 5 Discussions and Conclusions

Virtual environments provide a promising venue for design collaboration. The channels and tools in a virtual environment go beyond the traditional media. Users can easily create, modify and manipulate shareable virtual objects in real time. Communication can be made via chats, voice, video, sketch, 3D model, etc. However, there is room to be improved and there are some fundamental issues that need to be resolved. Some of these are related to our perception and cognition of work mode and collaboration. The default working scale in virtual environments is one-to-one. Designers, especially architects and town planners have been trained to work with different scales at different point of a project. In this chapter, we have addressed the issue of "scale" and its related matters such as presence, gesture and design communication content.

Gesture for design communication in virtual environments has not be broadly applied and emphasized. Gesture in Second Life (in virtual environments) provides another approach for communication. It has potentials in being applicable to design communication.

Protocol analysis has limitations and difficulties in analyzing non-verbal and non-storable design communication, like those gestures, when studying people collaboration in the physical world. However, in the virtual world users' actions can be captured in a log file. If the service provider can provide this log file, it will be a good resource to accompany the verbal protocol to investigate the designers' behaviors in a virtual environment.

## Acknowledgment

Special thanks to Dr Yinghsiu Huang at the Digital Design Department of the MingDao University Taiwan for his discussions and contributions as well as students at the Faculty of Architecture, Design & Planning of the University of Sydney Australia and the Digital Design Department of the MingDao University Taiwan who participated in the collaborative design projects and their contributions.

## References

Allen, T.J., Lee, D.M., Tushman, M.L.: R&D performance as a function of internal communication, project management, and the nature of work. IEEE Transactions on Engineering Management, EM 27, 2–12 (1980)

Bull, P.: Communication Under The Microscope: The Theory and Practice of Microanalysis, Routledge, Hove, East Sussex (2002)

Cross, N., Christiaans, H., Dorst, K.: Introduction: the Delft protocols workshop. In: Cross, N., et al. (eds.) Analysing Design Activity, pp. 1–14. John Wily & Son, Chichester (1996)

Cross, N., Cross, A.C.: Observation of teamwork and social processes in design. In: Cross, N., et al. (eds.) Analysing Design Activity, pp. 291–317. John Wily & Son, Chichester (1996)

Eastman, C.: On the analysis of intuitive design processes. In: Moore, G.T. (ed.) Emerging Methods in Environmental Design and Planning: Proceedings of the Design Methods Group first International Conference, Cambridge, MA, June 1968, pp. 21–37 (1970)

Friedman, D., Steed, A., Slater, M.: Spatial Social Behavior in Second Life. In: Proceedings of the 7th international Conference on Intelligent Virtual Agents, pp. 252–263. Springer, Heidelberg (2007)

Hewes, D.E.: A socio-egocentric model of group decision-making. In: Hirokawa, R.Y., Poole, M.S. (eds.) Communication and Group Decision-Making, pp. 265–291. Sage, Beberly Hills (1986)

Hewes, D.E.: Small group communication not ingluence dscision making: An amplication of socio-egocentric theory. In: Hirokawa, R.Y., Poole, M.S. (eds.) Communication and Group Decision Making, pp. 179–214. Sage, Thousand Oaks (1996)

Kavakli, M., Gero, J.S.: Strategic knowledge differences between an expert and a novice designer: an experimental study. In: Lindemann, U., Dummy (eds.) Human Behaviour in Design (2003)

Kvan, T.: Collaborative design: What is it? Automation in Construction 9(4), 409–415 (2000)

Levinson, S.C.: Pragmatics. Cambridge University Press, New York (1983)

Minneman, S.L.: The Social Construction of a Technical Reality: Empirical Studies of the Social Activity of Engineering Design Practice, PhD Thesis, Mechanical Engineering, Stanford University, Stanford (1991)

Olson, G.M., Olson, J.S.: Distance matters. Human -Computer Interaction 15(2/3), 130–178 (2000)

Oslon, G.M., Oslon, J.S., Carter, M.R., Storrosten, M.: Small group design meetings: an analysis of collaboration. Human -Computer Interaction 7(4), 347–374 (1992)

Salter, A., Gann, D.: Sources of ideas for innovation in engineering design. Research Policy 32(8), 1309–1324 (2002)

Sanchez-Vives, V., Slater, M.: From presence to consciousness. Nature 6, 8–15 (2005)

Sonnenwald, D.H.: Communication roles that support collaboration during the design process. Design Studies 17(3), 277–301 (1996)

Stempfle, J., Badke-Schaub, P.: Thinking in design teams - an analysis of team communication. Design Studies 23(5), 473–496 (2002)

Suwa, M., Purcell, T., Gero, J.S.: Macroscopic analysis of design processes based on a scheme for coding designers' cognitive actions. Design Studies 19(4), 455–483 (1998)

Suwa, M., Purcell, T., Gero, J.S.: Unexpected discoveries and s-invention of design requirements: important vehicles for a design process. Design Studies 21(6), 539–567 (2000)

Suwa, M., Tversky, B.: What do architects and students perceive in their design sketchs? A protocol analysis. Design Studies 18(4), 385–403 (1997)

Tang, H.H.: Exploring the Roles of Sketches and Knowledge in Design Process, PhD Thesis, Architecture, University of Sydney, Sydney (2002)

Tang, J.C.: Findings from observational studies of collaborative work. International Journal of Man-Machine Studies 34, 143–160 (1991)

Tsai, J.J.-H., Wang, X., Huang, Y.: Studying different scales of collaborative designs in real and virtual environments. In: ANZAScA2008, Newcastle, Australia, pp. 277–284 (2008)

Tsai, J.J.-H., Wang, X., Kan, J.W.T., Huang, Y.: Impacts of Different Scales on Design Collaboration in 3D Virtual Environments (2009) (working paper)

Tushman, M.L., Katz, R.: External communication and project performance: An investigation into the role of gatekeepers. Management Science 26(11), 1071–1085 (1980)

Vera, A.H., Kvan, T., West, R.L., Lai, S.: Expertise and collaborative design. In: CHI 1998 Conference Proceedings, Los Angeles, pp. 503–510 (1998)

Visser, W.: The function of gesture in an architectural design meeting. In: DTRS7 Design Meeting Protocols Proceedings, pp. 197–210 (2007)

Zolin, R., Hinds, P.J., Fruchter, R., Levitt, R.E.: Interpersonal trust in cross-functional, geographically distributed work: a longitudinal study. Information and Organization, 1–26 (2004)

van Someren, M.W., Barnard, Y.F., Sandberg, J.A.C.: The Think Aloud Method: A Practical Guide to Modelling Cognitive Processes. Academic Press, San Diego (1994)

# Design Coordination and Progress Monitoring during the Construction Phase

Feniosky Peña-Mora[1], Mani Golparvar-Fard[2], Zeeshan Aziz[3], and Seungjun Roh[2]

[1] Columbia University, Newyork, USA
[2] University of Illinois at Urbana-Champaign, USA
[3] University of Salford, UK

**Abstract.** Well-depicted baseline, systematic data collection, rigorous comparison of the as-planned and as-built progress and effective presentation of measured deviations are key ingredients for effective project control. Existing construction progress monitoring techniques are constrained by their reliance on ad-hoc approaches. This chapter presents advanced computer visualization and color and pattern coding techniques to compare the as-planned with the as-built construction progress. Key objective is to provide a method for automated collection, analysis and visual reporting of construction progress to facilitate the decision-making process. Construction progress is visualized by superimposing 3D Building Information Models (BIM) over construction photographs. The framework uses context-aware technologies to capture user's context and the need for information. A complementary 3D walkthrough environment provides users with an intuitive understanding of the construction progress using color coding and patterns.

**Keywords:** building information model, progress monitoring, time lapse photographs.

## 1 Introduction

A key objective during execution phase of a construction project is to achieve key project objectives within the specified budget and schedule. During project execution phase, key project stakeholders (e.g., owners, designers and contractors) frequently arrange meetings to discuss construction activities, monitor budget and schedule progress, discuss site safety and logistical concerns, update construction schedules, undertake on-site inspections and resolve field conflicts and drawing discrepancies. Accurate and rapid progress assessment provides an opportunity to understand the current performance of a project and help project managers make timely decisions if there is any deviation from the planned work. Thus, there is a need to design, implement, and maintain a systematic and comprehensive approach for progress monitoring to promptly identify, process and communicate discrepancies between the actual (as-built) and as-planned performances as early as possible. Existing progress measurement methods have limitations on accuracy

X. Wang & J.J.-H. Tsai (Eds.): Collaborative Design in Virtual Environments, ISCA 48, pp. 89–99.
springerlink.com 

and reliability due to ad-hoc measurements based on the personal experience of the project manager (Yoon et al., 2006). Also, various progress measurement methods are applied inconsistently without objective criteria on a project by project basis (Chin, 2004). In general, the lack of communication, clarity, coordination and organization is one of the most cited problems during construction project execution. Other limitations of existing progress monitoring methods as cited in the literature include: they are time consuming and data intensive (Navon and Sacks, 2007); the manual methods are prone to human errors. Other limitations are the use of non-systematic methods such as weighted milestones and budget-based monitoring (Meredith and Mantel, 2003), inability to represent multi-variable and spatial information (Kymell, 2008; Poku and Arditi, 2006). This chapter reviews emerging technologies in Section 2 that are being used for progress monitoring and presents two innovative approaches in Section 3 and 4 to construction progress monitoring that addresses some of the aforementioned limitations.

## 2 Emerging Technologies for Progress Monitoring

This section reviews several emerging technologies that have been applied for construction progress monitoring.

### 2.1 Tagging Technologies

Barcode and radio frequency identification (RFID) tags have been used to capture and transmit data from a tag embedded or attached to construction products (e.g., Kiziltas et al., 2008; Navon and Sacks, 2007). Such data can be used to capture construction progress. Unlike barcodes, RFID tags do not require line-of-sight, individual reading and direct contact (Kiziltas et al., 2008). Data can be dynamically updated on active RFID tags. Although RFIDs and barcodes potentially eliminate non-value adding tasks associated with project management processes, they require frequent installation and maintenance. Additionally, they cannot be attached to many types of construction components and they do not capture progress of partially installed components (Golparvar-Fard et al., 2009a).

### 2.2 Laser Scanners

Laser scanners have been used for construction quality control (Akinci et al., 2006), condition assessment (Gordon et al., 2004), component tracking (Bosche and Haas, 2008), automated data collection and progress monitoring (El-Omari and Moselhi, 2008). Key challenges in implementing Laser Scanning technologies on construction sites include discontinuity of the spatial information, portability, mixed pixel phenomenon (Kiziltas et al., 2008) as well as scanning range and sensor calibration. Laser scanners provide Cartesian coordinate information from the scanned scene. Processing significant amount of this information is time-consuming. Also, scan data do not carry any semantic information, such as which point belongs to what structural components. Working with this type of featureless data makes geometric reasoning based on this data tedious and error prone (Kiziltas et al., 2008).

### 2.3 Location Tracking Technologies

Such technologies can be used for capturing location-relevant progress monitoring data. Global Positioning System (GPS) requires a line-of-sight between the receiver and the satellite and is not suitable for location tracking in indoor environments. Behzadan et al. (2008) suggested using Wireless LAN (local area network) technology as a tracking technique for indoor locations, however, they also reported issues in accuracy and quality resulting from dynamic conditions on construction site. This requires regular calibration to maintain a high level of accuracy.

Other technologies such as wearable computers, mobile devices along with speech recognition and touch screens have also been used to capture construction site data electronically (Reinhardt et al., 2000). Implementation of aforementioned techniques on a construction site is often constrained by excessive costs, technology implementation related issues and additional workload resulting from technology implementation (Kiziltas et al., 2008).

## 3 Construction Progress Monitoring with 4D Simulation Model Overlaid on Time-Lapsed Photographs

This section presents an innovative construction progress monitoring approach using 3D/4D (3D + time) models as as-planned data repositories to facilitate accessing geometrical information, visualizing planned schedule and communicating progress. Such 3D/4D models provide a consistent visual base-line platform of as-planned information (Golparvar-Fard et al., 2009a) and help to visualize deviations between the as-planned and as-built progress. For collection, analysis and communication of as-built data, daily progress photographs were used to visually present daily work progress data.

The visualization model (see Figure 2) integrates the 4D model and time-lapsed photographs within an Augmented Reality (AR) environment where progress discrepancies are identified and visualized. Recent advances in digital photography and web camera technology means that image/video capture of construction progress is much more cost-effective and practical. Such time lapse photographs/videos are being used to create a photo log for dispute resolution/litigation purposes and recording field operations for multiple purposes including progress monitoring and reporting. Table 1 shows the advantages and drawbacks of such approaches.

The visualization process consists of a series of modules which results in color coded time-lapsed AR imageries. Figure 1 summarizes the information action-representation-environment perspectives for the proposed system. Raw data is collected from the as-planned and the as-built performance environments. The collected information represents product models, i.e., IFC 3D as-planned model

**Table 1.** Advantages and drawbacks of time-lapse photography and videotaping (Golparvar-Fard et al. 2009b)

| Advantages | Drawbacks |
|---|---|
| • Easy to obtain progress images<br>• Inexpensive<br>• Time-Lapse photography continuously records and yields benefits of filming without diminishing understanding of the operations that is recorded<br>• Easily understandable<br>• Provide more detailed and dependant information<br>• Making possible review and study by analysts, management, or other groups away from the construction site<br>• Suitable for progress monitoring and productivity analysis | • Distortion of images – make it challenging to superimpose images<br>• Show what is not obstructed by objects such as construction machinery or scaffoldings<br>• Show what is within range and view field<br>• Various illumination, shadows, weather and site conditions makes it difficult for image analysis<br>• Storage of digital photographs/ videos |

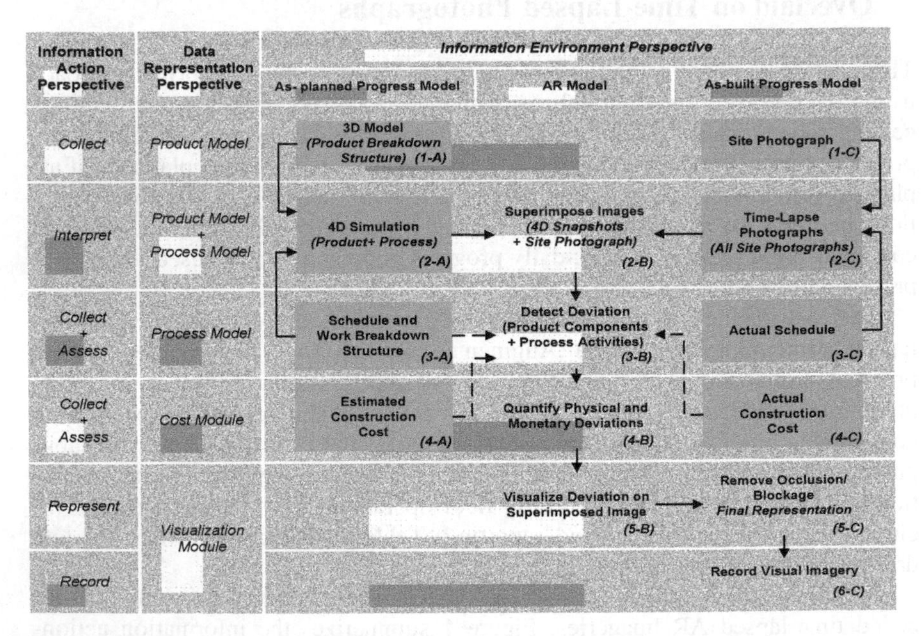


**Fig. 1.** Information Action-Representation-Environment perspectives for visualization of construction progress monitoring. (Golparvar-Fard et al. 2009b)

and site photographs (Figure 1: 1-A and 1-C), process models, i.e., working schedule and operation process (Figure 1: 3-A and 3-C) and cost modules, i.e., estimated and performed costs (Figure 1: 4-A and 4-C).

Collected information from these two environments is merged to produce a 4D as-planned simulation and time-lapsed photographs (Figure 1: 2-A and 2-C respectively). For any given time, the as-planned model is superimposed on the as-built performance model (i.e., site photograph) (Figure 1: 2-B) via proper registration of the 3D model and photograph coordinates. The superimposed imagery would allow discrepancy to be either manually or automatically detected and quantified (Figure 1-3-B and 1-4-B). At this stage, cost values are extracted from estimated and actual construction cost modules and are integrated with the system (Figure 2-4-A and 2-4-C). This would allow cost information required for Earned Value Analysis (EVA) to be derived. This information is appended to the known as-planned and as-built information and allows the budget spent, cost discrepancies and planned value to be assessed.

As the following step, a series of visualization techniques are applied to visualize EVA metrics (Figure 1: 5-B); i.e., according to the status of the progress, the as-planned model would be color-coded and superimposed on top of the photograph. Once the status of the progress is represented on the superimposed image, any occlusion and blockage caused by the superimposition should be removed. Therefore depth and perspective integrity of the virtual and actual environments is maintained (Figure 1: 5-6C). The final imageries are represented for decision making and are kept as a record for progress monitoring photo log. Figure 2 shows a color-coded superimposed image where the progress status is visualized. In this figure, on-schedule entities are represented in light-green entities, ahead of schedule entities in dark green, and behind-schedule entities in red color respectively. This reporting process is repeated for every coordination cycle where control actions are taken and the construction schedule is revised and updated by project participants. Once the camera is registered, the as-planned model (Figure 2-b) can be superimposed on the photograph. At this step, discrepancies between the as-planned model and the photograph (as-built) are easily identified (Figure 2-c). Given the deviations observed, EVA metrics are obtained (compared to schedule or cost information (Figure 2-f) and a color (depending on the progress status) is assigned to the planned model.

This preliminary method has shown that Augmented Reality environment can successfully represent progress monitoring information in forms of as-planned, as-built information along with their comparison in a holistic manner. The superimposed images retain the construction site information while the planned information along with the status of progress is enriching the contextual information within these photographs. The registration method gives the opportunity for image processing to be applied to specific regions within the photograph to assess the status of the progress based on material and shape recognition techniques. Color-coding metaphors give the end users of the superimposed photograph the opportunity of grasping progress status based only on a single representation form and could facilitate communication of progress status within a coordination meeting, allowing more time to be spent on control decision making. Moreover, preliminary results of

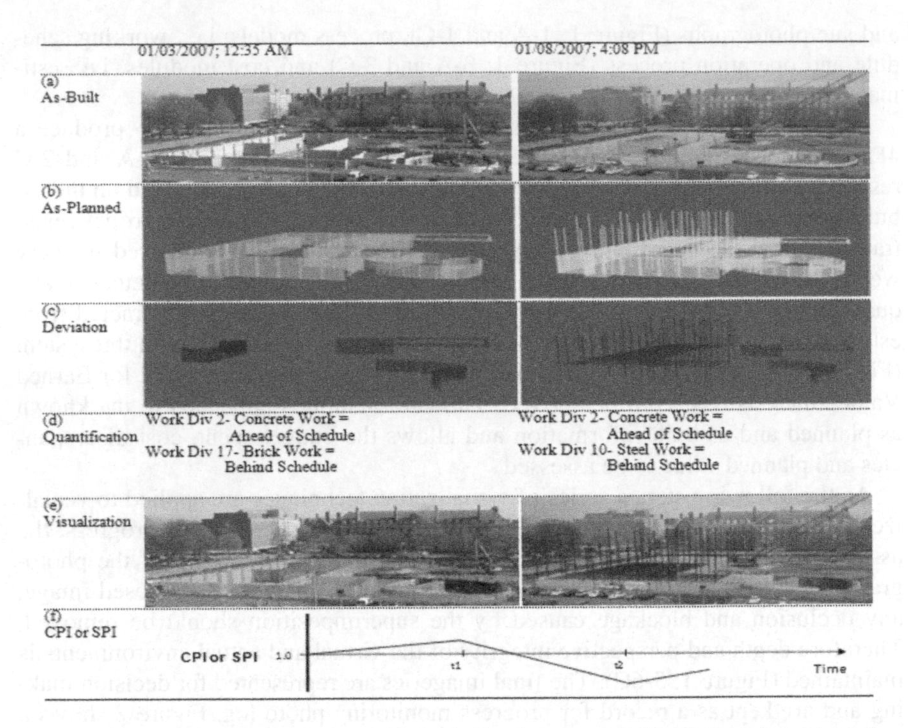


**Fig. 2.** Visualized monitoring report: (a) As-built photographs, (b) 4D snapshots, (c) color coded virtual components, (d) quantification of the deviation, (e) augmented photographs and (f) measured EVA performance metrics (Cost performance metric (CPI) and Schedule performance metric (SPI)) (Golparvar-Fard et al. 2009b)

applying feature detection technique preserves depth and perspective within superimposed photograph allowing a more realistic picture of the progress status to be represented. The overall methodology and reporting addresses issues related to data collection and reporting of a robust progress monitoring.

## 4 Interior Construction Progress Monitoring

The execution stage of a construction project can be divided into two key stages: shell and core construction (i.e. the basic structure and façade) and fit-outs (i.e. installation of building services such as mechanical, electrical, and plumbing (MEP) work). The fit-out stage is relatively complex involving a large number of specialty sub-contractors to install various building services. This makes interior construction progress monitoring relatively complex and approaches such as those involving photographs cannot be easily employed because of physical obstructions and requirements of a large number of cameras to capture multiple perspectives. Existing approaches are not adequate in providing spatial context and in representing complexities of interior components (Koo et al., 2007). This section presents

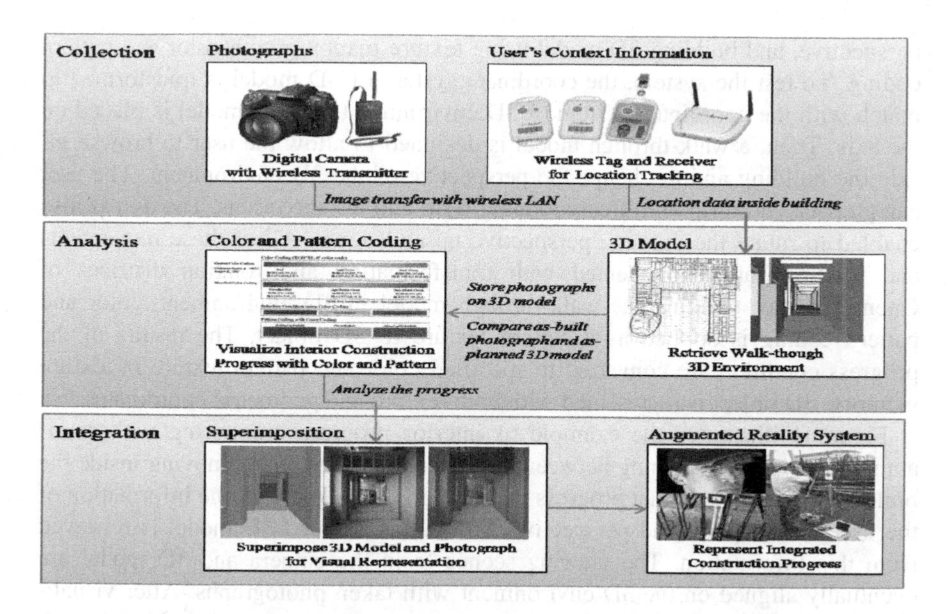


**Fig. 3.** System Processes of Interior Progress Monitoring (Roh, 2009)

application of advanced computer visualization techniques to visually represent interior progress by mapping building objects from as-built photos and as-planned building model.

Figure 3 presents a framework for interior construction progress monitoring comprising three key components of collection, analysis and integration. In the collection phase, as-built visual data is captured through digital photographs, while user context data is captured through different types of sensors (e.g., tags, wireless networking infrastructure, user location, etc.). Captured location data provides spatial information to allocate user's position in a walk-through model. As-built construction photographs are linked to a 3D environment in real-time to help build a walk-through model. Color and pattern coding are used for visualization of interior construction progress to analyze the progress of the elements in 3D model as compared with as-built photographs. Thus, the 3D model and photographs are superimposed using computer visualization technologies, and the results of interior construction progress are represented with Augmented Reality systems.

To develop a real-time 3D interior progress monitoring system, based on real time user location, viewpoint and perspective, as-built photos should be aligned and positioned in a 3D model. It also requires that users can easily browse 3D walk-through environment to find photographs in 3D model. OpenGL (Open Graphics Library) is used to define locations of 3D objects and photographs related with the viewing system. Figure 4 shows a prototype application for placing 3D model in walk-through 3D environment, transforming user's view point and perspective, and building 3D model using texture mapping with color and pattern coding. To test the system, the coordinate system of 3D model is transformed to match with the coordinate system of 3D environment and 3D model is placed on the axis. Then, a walk-through model is designed to allow the user to browse inside the building and to change the perspective of the 3D environment. The user can go back and forth with freely rotating right and left directions. The user is also enabled to rotate the viewing perspective up and down. All of these movements and rotations are implemented with transformation and rotation matrices of OpenGL. After building the walk-through model in 3D environment, color and pattern coding is applied as a pattern texture to 3D object. The results of the progress condition are converted to specified color and pattern texture in texture memory. 3D object is represented with both texture and geometric coordinates.

Figure 4 illustrates the example of interior progress monitoring with semi-automated camera matching between 3D model and the camera moving inside the building. It is assumed that progress photographs are taken with the information of the user's view point and perspective, and the as-planned 3D model is retrieved from this information. The viewing scenes from the camera and 3D model are eventually aligned on the 3D environment with taken photographs. After visualization system manually matches construction progress photographs with 3D model, the progress condition is represented with color and pattern coding on 3D model. For example, superimposed photograph in Figure 5 is performed based on as-planned progress information in color and pattern coding representation. In black and white printing, color coding is not identified, however pattern is still working to present interior construction progress.

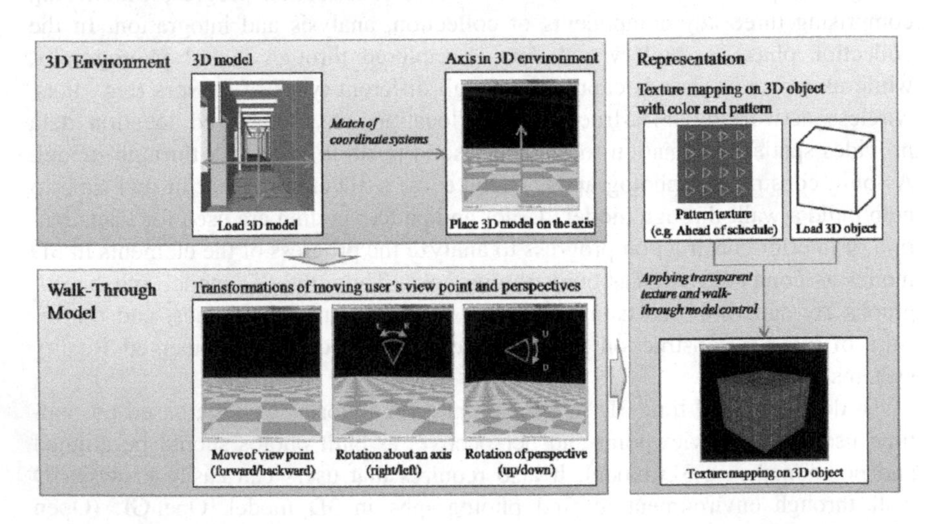


**Fig. 4.** Prototype Application of Interior Progress Monitoring (Roh et al., 2009)

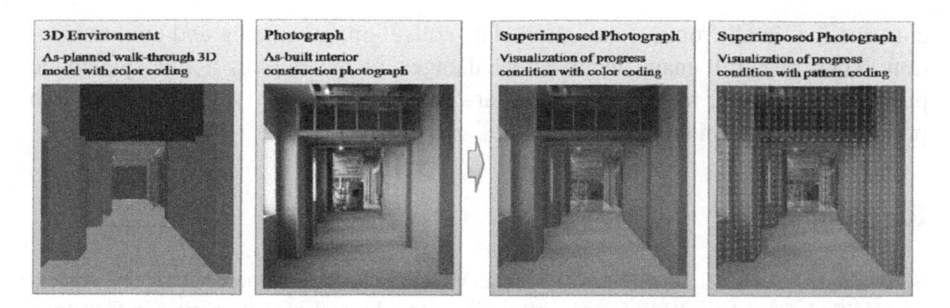


**Fig. 5.** Examples of Interior Progress Monitoring in 3D Environment (Roh et al., 2009)

Golparvar-Fard et al. (2007) reported other visualization techniques such as using colour quadrangle to jointly represent cost and schedule performance indices or using different color gradients to represent different levels of schedule deviations or floats. Figure 6 further illustrates one of such representations:

**Fig. 6.** Visual representation of overall project performance (Golparvar-Fard et al., 2007)

# 5 Discussion and Conclusions

Joint visualization of as-built and as-planned construction can enhance identification, processing and communication of progress discrepancies. To that end, we have proposed superimposition of BIM as-planned models over site photographs either taken from a fixed point of view (time-lapsed) or from different locations using location tracking techniques. Within such an Augmented Reality environment, progress photographs are registered in the virtual as-planned environment allowing a large unstructured collection of daily construction images to be sorted, interactively browsed and explored. In addition, geo-registered site imagery allowing a location-based image processing technique to be used and progress data to be automatically extracted. Such an integrated visualization can significantly facilitate contractor coordination and communication. Our preliminary results show perceived benefits and future potential enhancement of this new technology in construction, in all fronts of automatic data collection, processing and communication. There are still many technical challenges in developing a full systematic progress monitoring system and these are under exploration within the research project highlighted in this chapter.

## References


Behzadan, A., Aziz, Z., Anumba, C., Kamat, V.: Ubiquitous location tracking for context specific information delivery on construction sites. Journal of Automation in Construction 17(6), 737–748 (2008)

Bosche, F., Haas, C.T.: Automated retrieval of 3D CAD model objects in construction range images. Journal of Automation in Construction 17(4), 499–512 (2008)

El-Omari, S., Moselhi, O.: Integrating 3D laser scanning and photogrammetry for progress measurement of construction work. Journal of Automation in Construction 18(1), 1–9 (2008)

Golparvar-Fard, M., Peña-Mora, F., Savarese, S.: D4AR- A 4-Dimensional augmented reality model for automating construction progress monitoring data collection, processing and communication. Journal of Information Technology in Construction, Special issue in Next Generation Construction IT (in press)

Golparvar-Fard, M., Peña-Mora, F., Arboleda, C., Lee, S.: Visualization of construction progress monitoring with 4D simulation model overlaid on time-lapsed photographs. ASCE Journal of Computing in Civil Eng., Special Ed. on Graphical 3D Visualization in AEC (2009b) (in press)

Golparvar-Fard, M., Sridharan, A., Lee, S., Peña-Mora, F.: Visual representation of construction progress monitoring metrics on time-lapse photographs, Construction Management and Economics. In: 25th Anniversary Conference, University of Reading, UK (2007)

Gordon, S., Lichti, D., Stewart, M., Franke, J.: Modelling point clouds for precise structural deformation measurement. In: XXth ISPRS Congress, Istanbul, Turkey (2004)

Kiziltas, S., Akinci, B., Ergen, E., Tang, P.: Technological assessment and process implications of field data capture technologies for construction and facility/infrastructure management. ITcon, Special Issue Sensors in Construction and Infrastructure Management 13, 134–154 (2008)

Koo, B., Fischer, M., Kunzc, J.: A formal identification and re-sequencing process for developing sequencing alternatives in CPM schedules. Automation in Construction 17(1), 75–89 (2007)

Kymell, W.: Building information modeling: planning and managing construction projects with 4D CAD and simulations. McGraw-Hill, New York (2008)

Meredith, J., Mantel, S.: Project Management: A Managerial Approach, 5th edn. Wiley, Chichester (2003)

Navon, R., Sacks, R.: Assessing research in automated project performance control (APPC). Journal of Automation in Construction 16(4), 474–484 (2007)

Poku, S., Arditi, D.: Construction scheduling and progress control using geographical information systems. Journal of Computing in Civil Engineering 20(5), 351–360 (2006)

Reinhardt, J., Garrett, J., Scherer, J.: The preliminary design of a wearable computer for supporting construction progress monitoring, Internationales, Kolloquium über die Anwendung der Informatik und der Mathematik. Architektur und Bauwesen, Weimar, Germany (2000)

Roh, S., Peña-Mora, F., Golparvar-Fard, M., Han, S.: Visualization Application for Interior Progress Monitoring. 3D Environment, 2009 Construction Research Congress (CRC), Seattle, Washington, April 5-7 (2009)

Yoon, S.W., Chin, S., Kim, Y.S., Kwon, S.W.: An application model of RFID technology on progress measurement and management of construction works. In: ISARC 2006, pp. 779–783 (2006)

# Part IV
# How Designers Design in Collaborative Virtual Environments

Impact of Collaborative Virtual Environments on Design Process
Nobuyoshi Yabuki (Osaka University)

A Pedagogical Approach to Exploring Place and Interaction Design in Collaborative Virtual Environments
Ning Gu and Kathryn Merrick
(University of Newcastle & University of New South Wales)

Sketch That Scene for Me and Meet Me in Cyberspace
Ellen Yi-Luen Do (Georgia Institute of Technology)

A Hybrid Direct Visual Editing Method for Architectural Massing Study in Virtual Environments
Jian Chen (University of Southern Mississippi)

# Part IV
# How Designers Design in Collaborative Virtual Environments

# Impact of Collaborative Virtual Environments on Design Process

Nobuyoshi Yabuki

Osaka University, Japan

**Abstract.** Collaborative Virtual Environment (CVE) is a form of environments where multiple users, whether remote or not, can collaboratively develop and control virtual 3D models, using advanced information and communication technology (ICT). CVE is expected to have significant impacts on design and construction of civil and built environments, as the following: CVE will contribute to improve the quality of civil and built environments, and decrease development period and project cost. CVE is expected to design more environment-friendly, green buildings and civil infrastructure by analyzing environmental aspects in the early stage of design. More creative design is also expected in CVE by the collaborative atmosphere among heterogeneous experts. CVE will contribute to obtaining social acceptance of design from citizens and stakeholders through better presentation and communication. CVE is expected to give significant impact on the reform of business processes of design and construction from design-bid-build to design-build.

**Keywords:** concurrent engineering, product model, project process, design-build, integration.

## 1 Introduction

To design civil and building environments is a creative, complicated and advanced activity. One expert usually cannot design the entire thing and many experts in various disciplines participate in the design process (Luo and Dias, 2004). The traditional design adopts a methodology that divides the design object into foundations, structures, facilities, aesthetics, etc. and that classifies design as preliminary, detail, construction, etc. Participants in designing civil and built environments include owners, architects, structural engineers, foundation engineers, facility engineers, construction engineers, governmental officers, other stakeholders, etc. Each expert executes his or her own task of the discipline and sends the results to the owner or other experts on the basis of contracts. This procedure is error-prone and tends to take much time because other experts have to re-enter the data generated by previous experts into their computers. If a problem occurs in the later phase, designer may have to return to former phase for modification, ending up

X. Wang & J.J.-H. Tsai (Eds.): Collaborative Design in Virtual Environments, ISCA 48, pp. 103–110.
springerlink.com 

with significant cost increase. Moreover, knowledge and experience in the downstream phase may not be utilized in the upstream phase, which could hinder the opportunity of modifying and enhancing previous design for the future.

The environment and energy issue is one of the most crucial items to tackle in the 21st century. Civil and built environments spend a large amount of fossil fuels and emit much carbon dioxide. In the design of buildings and infrastructure, environmental aspects should be investigated and incorporated in the early stage of design, especially for designing energy-efficient environments. However, environmental aspects are usually analyzed in the later stage and treated like adjuncts rather than key roles.

Design and construction of civil and built environments often confront claims by citizens and other stakeholders, mainly due to misunderstanding during the design phase. Such troubles may stop the project and could lead to a severe damage to the project developers.

## 2 Concurrent Engineering

In mechanical design field, 3D CAD systems have begun to be used since 1970s and CAD was merged with Computer-Aided Manufacturing (CAM) and Computer-Aided Engineering (CAE), which lead to Computer-Integrated Manufacturing (CIM). In CIM of automobiles, bodies and main parts are designed using CAD, then the geometric and property data is transferred to structural analysis software packages based on the finite element method (FEM), and various simulations are performed. The design is then modified based on the result of simulations. After the design is complete, steel plates are cut and drilled using numerical control (NC). Since software packages such as CAD, FEM, simulations, NC, etc. were developed by different companies and researchers, the interoperability of data had become an important issue. In early 1980s, the object-oriented paradigm was born, and university researchers and industry engineers began to develop common data models for representing products, i.e., product models. Some of such efforts have become parts of Standard of the Exchange of Product model data (STEP) of International Organization for Standardization (ISO) (Ota and Jelinek, 2004).

As the model-based method of design and manufacturing spread in industry, the concept of concurrent engineering emerged. In concurrent engineering, an object to-be-designed is divided into several parts, or is divided into several design tasks, which are executed by each expert or group of experts separately and concurrently. Then, the experts meet and merge the results of the design. Based on the review of the design, they execute the next tasks separately, and this process is repeated until the design is done. Concurrent engineering has been expected to get the design done faster and in a shorter time than the traditional procedural design. Further, since multiple experts in heterogeneous disciplines gather, discuss freely and generate new ideas, concurrent engineering has been expected to generate more creative design (Vivasqua and de Souza, 2004).

However, while experts are doing their own tasks, they do not know what others are doing. Thus, when they show their results each other at the meeting, problems such as conflicts of design objects usually occur and they have to solve the problems by discussion. Sometimes some of the design results may be wasted. As values, knowledge, and terms of experts in various disciplines are often different, they tend to argue without full understanding, have difficulty in communication, and cumulate frustration. Although they should discuss with other experts during the parallel work period if they encounter a trouble or waver in decision-making, they tend not to. Actually, although concurrent engineering looked promising then, it is not so easy in practice.

## 3 Integration

Design of civil and built environments varies according to the phase from preliminary design, detailed design, to construction design. Since civil and architectural structures are built in natural or urban environment, all the conditions or specifications cannot be given a priori at the beginning of design. Much information is collected by surveying, site investigation, etc. as the design proceeds. Thus, design is often modified or sometimes forced to change itself and the amount of information of the design becomes larger and larger in the course of development.

In the design of civil and built environments, requirements such as safety, serviceability for different types of users, aesthetics, impact for nature and environment, cost, schedule, etc., must be investigated and satisfied. In addition, adjunct facilities and machines have to be designed. Therefore, many experts in heterogeneous disciplines execute investigation in each area. The amount of work and the level of accuracy are different according to the design stage. Most of the design tasks are done using computer software packages. As in the mechanical design area, the interoperability of data among software packages was low, development of product models for buildings started in 1990s. Now the de facto standard of building product models is Industry Foundation Classes (IFC).

Since around 2005, the term Building Information Modeling (BIM) has been spread in building industry. BIM has not been clearly defined yet, but it means a technology for executing design of buildings and civil infrastructure quickly and efficiently by exchanging and sharing design information based on the product models such as IFC that integrates heterogeneous application software systems. The International Alliance for Interoperability, which is in the process of changing its own name to buildingSMART, has organized events called BIM-Storm recently. In BIM-Storm, each participating group consists of several design experts in heterogeneous disciplines such as architecture, structural design, facility management, construction management, etc. The groups compete to design an assigned building by exchanging and sharing design information based on IFC via the Internet in only a few days. In Japan, a similar event called Build Live Tokyo 2009 was organized and six groups participated in it. They designed a virtual building in a certain place in Tokyo within 48 hours. Many experts in building industry are trying to change the design process significantly by the concept of BIM.

## 4 Project Processes

The process of design and construction of civil and built environments, usually of single part production, is considerably different from that of products manufactured and mass-produced at factories. Especially, as for public civil infrastructure, design and construction are separately ordered by governmental, municipal and other public agencies, and the same private company is not able to receive both design and construction orders in principle. Because if one company performs both design and construction, the structure may be designed unnecessarily larger or more luxurious so that the construction cost becomes higher than normal, which is a serious concern for clients or owners. Although the design and construction separate ordering method, i.e., design-bid-build is effective for preventing corruption, it is very difficult to adopt the concurrent engineering method among design consultants and general contractors unlike manufacturing industry. Thus, consultants' design is not usually reflected by knowledge and experience of contractors, which may lead to various problems during construction. Further, if consultants know more new construction methods created by contractors, better and cost reduced design may be possible. Sharing and using knowledge and experience of construction by contractors is increasingly important and effective in the design phase.

In the United States and some other countries, Construction Manager (CM) with a wealth of knowledge and experience in design, construction, and management works among owners, design consultants, and contractors for managing a construction project smoothly from planning to complete of construction. In Japan, although CM has not been widely adopted yet, design-build method, which is commonly used in private building projects, is now under consideration for applying to public civil engineering works. Furthermore, new project methods such as Build-Operate-Transfer (BOT), Private Finance Initiative (PFI), Public Private Partnership (PPP) are investigated for public works. These new project methods are expected to break the limitations of the traditional design and construction separate ordering method.

## 5 Social Acceptance of Design

Design of civil and built environments is not satisfactory if it is not accepted by the society yet. The design becomes realistic only if it is understood and accepted by citizens and other stakeholders, i.e., society. However, even if the designed project is thought to be understood by the citizens and stakeholders, once the construction is started, they may think that the situation is different from what they thought. Because professional documents such as 2D drawings and reports describing the project tend to be hard to fully understand for novice people. The gap between their understanding and the reality could lead to a serious social trouble or problem that may end up re-design, postponement or cancellation of works.

Thus, when government, public agencies, or private developers perform urban planning, civil and building engineering projects, they often develop 3D computer graphics (CG) models representing the urban areas, natural environments, etc. and present them to the citizens and other stakeholders at public hearings, using virtual reality (VR) software. Figure 1 shows snapshots of VR presentation of light rail transit (LRT) project planned in Sakai, Osaka, Japan (Yabuki et al., 2009).

At public hearings or explanation meetings, VR tends to be used just for presenting planned or designed projects in an easy-to-understand manner. However, Collaborative Virtual Environments (CVE) could change presentation and discussions at public hearings so that developers can modify the 3D VR model according to the request of stakeholders and present what-if scenarios interactively

**Fig. 1.** Snapshots from presentation of the LRT project in Sakai, Osaka, Japan.

# 6 Collaborative Virtual Environments

Computer technology is fast evolving. Concurrent engineering, CIM, CSCW, etc. develop into CVE by the evolution of 3D/4D CAD and VR technologies. CVE is a form of environments where multiple users, whether remote or not, can develop and control virtual 3D models with immersive feeling, communicating and collaborating each other. For example, an immersive VR dome facility named CyberDome of Panasonic Electric Works has a dome of 8.5m in diameter, and 18 projectors are used to display 3D stereoscopic VR images and videos. Thirty people can participate in the dome simultaneously.

In the research field, not only visualization but also multi-modal CVE systems are being developed. Multi modal includes not only vision, smell, taste, listening, and touch but also input device such as 3D mouse, voice recognition, haptic feedback (Di Fiore et al., 2004).

The central part of CVE is a product model, which enables users to share and exchange data of products, processes, organizations, etc. Utilizing CVE, owners, designers, contractors, and other experts will collaboratively share and exchange design data. They will concurrently execute and compare a number of design alternatives, and will explain the plan and design to citizens and other stakeholders and negotiate immediately with them, considering the impact to cost and other conditions.

# 7 Impacts of CVE

CVE is expected to give significant impacts on design and construction of civil and built environments. The following sub sections describe impacts of CVE on design and construction.

## 7.1 Quality, Time and Cost

CVE will contribute to decrease of mistakes in design and construction, which will be effective for improvement of quality and decrease of accidents. Many mistakes in design and construction tend to occur due to misunderstanding in generating and reading 2D drawings and in communication among participants in design. CVE gives 3D VR world that designers and engineers can see and feel immersed in it, so that they can detect mistakes easily.

As participants for design, even if they are remote, can communicate with ease in CVE by video conferencing, decisions can be made more quickly and thus, wasted design and investigation will be decreased, which will reduce the period of design and construction and project cost.

## 7.2 Environmental Aspects

CVE is expected to contribute to realize the design which environmental aspects are more emphasized. As mentioned before, in current design of buildings and structures, most environmental analyses are done after basic design decisions are already made. It has been quite difficult to change the design significantly for improving energy saving, for decreasing the emission of carbon dioxide, and for giving less impact to environment. In CVE, environmental engineering experts will be able to participate in design from the early stage and suggest better design.

## 7.3 Creative Design

CVE is also expected to contribute to generate more creative design. Basic design done by an architect alone usually has limitations, while collaboration among several heterogeneous experts tends to generate more original ideas. Interesting and creative ideas are likely to be blocked in mind of designers because such ideas look unrealistic or too expensive or fragile. However, if experts in other fields get to know the idea, they can investigate the feasibility immediately in CVE.

## 7.4 Social Acceptance

CVE will contribute to better presentation and communication at public hearings and mutual understanding by citizens, stakeholders, designers, engineers, and owners. As hardware and software of CVE evolve continuously, public agencies and designers will soon be able to perform what-if analysis corresponding to the request from participants at public hearings and compare multiple virtual worlds concurrently. Current public hearings generally take place at halls. Future public hearings will take place more and more virtually and remotely via the Internet in the CVE, where citizens will ask questions, give opinions, criticize some points, etc. and design experts respond to them.

It takes considerable time and cost to develop a 3D VR urban model covering wide area. CVE does not necessarily mean 3D VR model but it may be a mixture of real and virtual worlds such as Mixed Reality (MR) and Augmented Reality (AR). MR/AR will be an effective means to realize CVE.

## 7.5 Business Process Reform

CVE is expected to give significant impact on the reform of business processes of design and construction of civil infrastructure and buildings. Current design-bid-build method is not collaborative. Each project participant does his or her own task alone based on the contract and specification. This process is approaching the limits of its performance. On the other hand, CVE will encourage project participants to meet and communicate with each other more than now. They can share 3D models of civil and built environments and investigate them with their software without much frustration. Then, they will realize what hinders the design process is not the technology any more but it is the business process, which is based on many, rigid, separate contracts.

New technology such as CVE and reform of business process are like two wheels of an automobile, which must work together inseparably. Old business process should be reformed so that new information and communication technology (ICT) will use its abilities to the full.

The author feels that design-build contract will be one of the most promising business processes for future public infrastructure design and construction, considering the affinity for CVE. Corruption in design-build must be banned. The third party's review and suggestions may be a solution to regulate and control the design utility.

We should identify problems and concerns in promoting CVE and business process reform and solve them for sustainable development and better quality of life.

# 8 Conclusion

In this chapter, current design and engineering processes were reviewed and issues and problems in design and construction of civil and build environments were identified. Current innovation in ICT and efforts in application of advanced VR and integration technologies evolve the traditional design environments into CVE. CVE is expected to give the following significant impacts on design and construction of civil and built environments.

- CVE will contribute to improve the quality of civil and built environments, decrease development period and project cost.
- CVE is expected to design more environment-friendly, green buildings and civil infrastructure by incorporating environmental analysis in the early stage of design.
- More creative design is also expected in CVE by the collaborative atmosphere among heterogeneous experts.
- CVE will contribute to obtaining social acceptance of design from citizens and stakeholders through better presentation.
- CVE is expected to give significant impact on the reform of business processes of design and construction from design-bid-build to design-build.

# References


Di Fiore, F., Vandoren, P., Van Reeth, F.: Multimodal interaction in a collaborative virtual brainstorming environment. In: Luo, Y. (ed.) Cooperative Design, Visualization, and Engineering, pp. 47–60. Springer, Berlin (2004)

Luo, Y., Dias, J.M.: Development of a cooperative integration system for AEC design. In: Luo, Y. (ed.) Cooperative Design, Visualization, and Engineering, pp. 1–11. Springer, Berlin (2004)

Ota, M., Jelinek, I.: The method of unified internet-based communication for manufacturing companies. In: Luo, Y. (ed.) Cooperative Design, Visualization, and Engineering, pp. 133–140. Springer, Berlin (2004)

Vivasqua, A.S., de Souza, J.M.: Fostering creativity in cooperative design. In: Luo, Y. (ed.) Cooperative Design, Visualization, and Engineering, pp. 115–122. Springer, Berlin (2004)

Yabuki, N., Kawaguchi, T., Fukuda, T.: A development methodology of virtual reality models for urban light rail transit projects. In: Caldas, C.H., O'Brien, W. (eds.) Proceedings of 2009 International Workshop on Computing in Civil Engineering, pp. 495–502. ASCE, Austin (2009)

# A Pedagogical Approach to Exploring Place and Interaction Design in Collaborative Virtual Environments

Ning Gu[1] and Kathryn Merrick[2]

[1] The University of Newcastle, Australia
[2] The University of New South Wales, Australia

**Abstract.** In most design schools, collaborative virtual environments (CVEs) are traditionally perceived and used as a computer-aided design (CAD) tool for 3D modelling and simulation. However, the design in virtual worlds can also be a stand-alone design subject that considers virtual worlds as a novel design environment for exploring place and interaction design, and not merely a technical tool for supporting design simulation. This chapter reports on our experience of teaching the design of virtual worlds as a design subject, and discusses the principles for designing interactive virtual worlds. We conclude by identifying future directions for designing and learning in virtual worlds.

**Keywords:** place design, interaction design, design education.

## 1 Introduction

There have been significant changes in design curricula to accommodate new demands, opportunities, processes and the potential provided by collaborative virtual environments (Kvan et al, 2004). Nevertheless, these environments are traditionally perceived and used as a computer-aided design tool for 3D modelling and simulation. This is because the early development of virtual worlds has been closely related to architectural design through the use of the place metaphor. By applying this metaphor, virtual worlds can inherit many of the characteristics from real-world architecture. However, more recent virtual worlds can go beyond imitating the physical world but still focus on accommodating human activities. They can support interactions – such as remote collaboration, adaptive and intelligent environments – that are not readily available in physical environments.

This chapter presents two case studies where the design of virtual worlds was taught as a design subject. These two cases consider virtual worlds as an alternative kind of environment for exploring place and interaction design, and not just as technical tools for supporting design simulation. The first case study focuses on teaching place design while the second one focuses on teaching interaction design.

X. Wang & J.J.-H. Tsai (Eds.): Collaborative Design in Virtual Environments, ISCA 48, pp. 111–120.
springerlink.com 

## 2 Case Study 1: Place Design

The place metaphor provides a way to understand virtual world layout, virtual object design and the issues associated with navigation (Champion and Dave, 2002; Kalay, 2004). However, to explore the full potentials of virtual worlds, designers need to think beyond the principles of physical architecture. The first case study is a studio-based course that encourages design students to consider virtual world design as 'alternative place' design and push the boundaries of place design conventions.

### 2.1 A Collaborative Design Studio: Course Overview

This course was established in August 2008 as the result of an on-going collaboration between the University of Newcastle, Australia and Rangsit University, Thailand. *NU-Genesis*, a virtual island in *Second Life* (Linden, 2009), was set up as the location of the studio. The aim of this studio was for students to (1) develop an understanding and hands-on experience of designing virtual worlds that extend conventional place design, and (2) understand and develop the essential skills of collaborative design and modelling in virtual worlds.

Students were first introduced to design principles for virtual worlds and remote collaboration skills. Next, students were guided to inhabit and critically assess a wide variety of virtual places in *Second Life*. Students from both universities collaborated remotely over a period of five weeks on a design project titled *Virtual Home*. The design brief required each group to design and implement a place in *Second Life* that demonstrates their concept of a virtual home and challenges the perceptions of a physical home. This builds on work done by students to develop a physical home in an earlier, traditional architectural studio.

### 2.2 Elements of Alternative Place Design

Four elements of alternative place design were explored in this course: zoning principles for virtual sites, design principles for individual buildings, tools to support design, and design approaches for virtual worlds. In each case, students were encouraged to think beyond the limitations of physical environments to make novel use of the available tools and develop novel solutions.

#### 2.2.1 Zoning Principles

Due to the technical limitations of virtual worlds, such as the number of objects that can be built in certain areas, there are often conflicts between the availability of buildable surfaces and the number of users. To plan and divide the *NU-Genesis* island for the collaborative project, an in-class design competition was conducted. The winning proposal excelled in its novel concept of the *Three Worlds* layout, and was adopted for the zoning development of the virtual island. The winning design incorporated the sites in a vertical structure on the island. Designs could utilise three different layers: underwater, on the ground or in the sky. As a result, the island was used to its full capacity. The *Three Worlds* layout also provided many unusual sites to enable the emergence of innovative design solutions in the collaborative project. Many groups were interested in selecting an unusual site, which they are unlikely to confront in a conventional architectural studio.

### 2.2.2 Design Principles

Alternative place design can be understood from the following two perspectives:

**Degree of realism in form:** Place designs applying dominantly simulated real-world forms are classified as 'realistic'. Designs adopting mainly forms that are imaginative, are classified as 'non-realistic'. Finally, designs that use a combination of both are classified as 'semi-realistic'.

**Degree of abstractness in concept:** Place designs having a profound meaning or concept behind their implementation is classified as having a high degree of abstraction.

Six selected designs that differ in their realism in form and abstractness in concept are shown in Figure 1: (1) *Sky Garden* (sky site): Explores the idea of a virtual home as a series of relaxing gardens. This design is relatively realistic in form; (2) *Archi-Bio* (ground site): Inspired by bio-mechanisms, this design transforms dynamic and growing attributes into a virtual home in *Second Life*; (3) *Metamorphosis* (underwater site): This virtual home design emulates different levels of sub-consciousness through the creation of ambient environments that depict different 'emotions'. This design has a high degree of conceptual abstraction; (4) *Floating Cubes* (sky site): Represents a home as a series of floating cubes that shift the occupants from one activity to another and from one mind-set to another; (5) *Zero Gravity* (sky site): Virtual worlds have no physical constraints such as gravity, but still support various activities. This design uses zero-gravity as the design trigger to challenge the constraint of gravity; and (6) *})i({*(underwater site): The virtual home design here is a place of communication inspired by poetry.

**Fig. 1.** Six selected Virtual Home designs.

Non-realistic and abstract designs represent a novel approach to place design and break from the conventional 'home' with innovative and challenging solutions. They also often lead to more interesting outcomes and encourage students to explore different design possibilities during the collaborative process, rather than repeating what they have already learnt in the conventional architecture studios.

### 2.2.3 Design Support and Tools

Students were able to explore and adopt a wide range of *Second Life* tools to assist their design and collaborative activities. These tools frequently differ from traditional design tools.

Individual identity appears to be an essential factor during design collaboration. Students not only spent a considerable amount of time customising their avatars to reinforce their virtual identities, but also used avatars as reference points when referring to design elements. For example, they often made statements such as "the red column next to ME". *Second Life* can enable students to become 'immersed in the experience' of interacting with design representations. This sense of immersion is defined as the level of fidelity that collaborative virtual environments provide to the user's senses (Narayan, 2005), which can be enhanced by the use of avatars.

*Second Life* supports a parametric modelling method by providing a set of basic 3D models (cubes, spheres, triangular prisms and so on). Certain parameters of these models can be adjusted by designers to make more complex shapes. First-person view and third-person view are supported during modeling so *Second Life* well supports the understanding of the spatial arrangement of the design elements, and the development of student's spatial abilities.

Some groups demonstrated a very high level of competency in applying different features of *Second Life* for different design phases. For example, in the *Archi-Bio* project, students strategically used different feature of *Second Life* to develop their design from an initial inspirational concept, shown in Figure 2(a); through to the abstract 3D models that assisted their conceptual development shown in Figure 2(b); and their final detailed implementation of the virtual home design, shown in Figure 2(c).

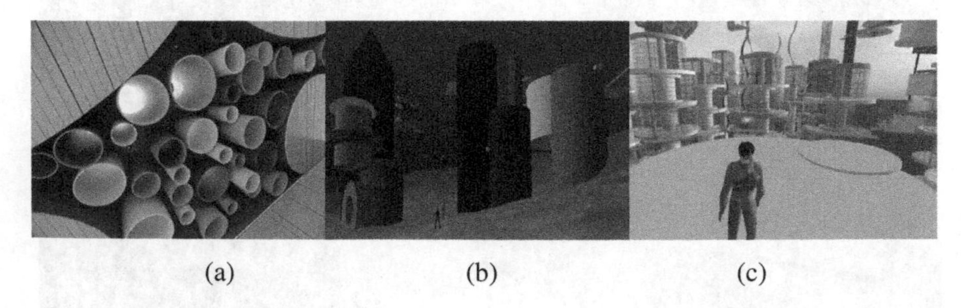

(a) (b) (c)

**Fig. 2.** The design development of the 'Archi-Bio' project in Second Life.

### 2.2.4 Design Approaches

We observed two different design approaches emerge from the design studio. The first one is the form-based approach where students start by exploring interesting forms, then adopt or sometimes even invent a concept afterwards. *Sky Garden* and *Floating Cubes* are among the designs that followed this approach.

The second approach is the concept-based approach, in which students firstly explore, develop and agree on certain concepts for the virtual place design, and then realise the concepts through 3D models.*})i({*, *Metamorphosis* and *Zero Gravity* are among the designs that adopted this approach.

## 2.3 Lessons Learned

A number of lessons can be learned from this case study. Firstly, although the parametric modelling tools provided by *Second Life* enabled students to start designing from the very early conceptual stage using basic geometric forms, this proved challenging and inadequate for some students. Students commented that they often had to sketch the design on papers in order to understand the overall design concept and layout, prior to modelling in *Second Life*.

The two design methodologies adopted by students also had different outcomes. Groups using the form-based approach often produced certain design solutions quickly then moved on to detailed design and documentation, as their design collaboration began with form-making and detailed object modelling. In contrast, groups using the concept-based approach frequently progressed slowly, especially in the early stage of the collaboration, compared to the groups that adopted the form-based approach. However, their design outcomes were often sophisticated and interesting, if they could successfully reach a shared understanding of the concepts and implement them afterwards.

Finally, in order for students to understand the relevance of the course to their own design disciplines, students need to be made aware of the importance of design activities set in a new context. They also need to understand how opportunities to challenge design boundaries are important extensions of their long-term design skills and career development.

# 3 Case Study 2: Interaction Design

Another unique property of virtual worlds is the capacity for the terrain, architecture, objects or computer controlled characters to be dynamic and interactive, adapting and responding to the actions of their inhabitants (Maher and Merrick, 2005). Teaching interactive environments is thus another key aspect of design education in collaborative virtual environments. This section presents a case study of the interaction design component of a postgraduate course on designing virtual worlds.

## 3.1 Designing Virtual Worlds: Course Overview

This course is offered to postgraduate students in design computing and digital media at the University of Sydney, Australia. The aim of the course is to introduce the basic techniques for designing and building virtual worlds and give students sufficient knowledge and hands-on experiences with software tools, to design and implement interactive, 3D virtual spaces. The focus is on the connection between the human activities supported by virtual worlds – including collaborative, entertainment, commercial, educational, design, defence and health applications – and the architecture required to support interactive, functional, multiuser environments. The course emphasises communication as a generic attribute, with students encouraged to identify, criticise and utilise design concepts for virtual worlds through individual and group tasks.

The course includes theoretical and practical components covering both place and interaction design. However in this section we will focus on the interaction design component. Two tutorials and one design project were dedicated to interaction design. *Second Life* is used as the learning platform for tutorials and projects.

## 3.2 Elements of Interaction Design

In the context of interaction design, the designed elements of virtual worlds can be broken down into three broad categories: (1) terrain or architecture, (2) characters, and (3) objects or artefacts (Duggan, 2007; Meigs, 2003). Unlike the physical world, where behaviour is usually attributed to characters (humans and animals), or sometimes objects or appliances, in virtual worlds behaviour can be attributed to any element of the environment. For examples, rooms may be able to change size or shape for different activities or furniture may be able to rearrange or reconfigure itself. A number of tools and techniques are available to achieve interactivity.

### 3.2.1 Interaction Design Tools

Interaction design tools can be divided into four main categories:

**Function libraries** are a short list of triggers and commands that may be performed by a virtual object. Triggers specify the condition that must be fulfilled for a command to be carried out. For example an object may rotate (command) when it is bumped (trigger). Several triggers and commands may be used together to achieve more complex behaviour.

**Scripting languages** define a syntax for programming world elements with complex sequences of commands. Scripting languages often incorporate a function library, but are more expressive than function libraries alone. For example they may allow the design of world elements that can make decisions or remember past events.

**Application Programming Interfaces** (APIs) are gateways through which external computer software can communicate with elements of a virtual world. External software can be written in an industrial strength programming language such as Java or C/C++. These languages are generally more powerful and expressive than scripting languages and can be connected to further third party tools such as databases or web pages.

**Open source client software** is a more recent means by which users can directly modify the virtual world. Designers can modify the world program itself, rather than just connecting their own external programs. This provides a powerful new tool for interaction design.

In this course students focused on learning the Linden Scripting Language (LSL) in practical sessions. This provided a good middle ground between the complexity of a full strength programming language and simple function library systems.

### 3.2.2 Interaction Design Techniques

Tools such as scripting languages and programming languages allow the designer to embed complex behavioural algorithms in world elements. These algorithms can range from simple artificial intelligence (AI) approaches (Russel and Norvig, 1995) to complex machine learning algorithms or cognitive agents. In this course, students were given a brief overview of three AI techniques – state machines, rule-based systems and agents (Wooldridge 2000) – but made aware that a much larger suite of techniques are available. These techniques are commonly used in games programming and thus relevant to interaction design in virtual worlds (Baille-De Byl, 2004; Maher and Gero, 2002; Maher and Merrick, 2005).

Students first experimented with these AI techniques in two tutorials. The theme for the first tutorial on state machines and rule-based AI was *Curious Places*. Students were asked to design interactive furniture with behaviours that had characteristics of human curiosity. Some of the results are shown in Figure 3(a). These include an array of cameras that are curious about inhabitants of the room and could track their movements; a chair that increases in height to give the audience a better view of interesting presentations; and a large hammer that swings to boot out avatars giving boring presentations.

In the second tutorial students experimented with implementing simple agent models. Each student was provided with a virtual sheep as shown in Figure 3(b).

**Fig. 3.** (a) 'Curious Furniture' in the virtual classroom, displaying behaviours that exhibit characteristics of human curiosity; (b) intelligent 'pet sheep'.

The sheep contained script for a simple agent framework. Students then had to implement rules within the agent framework to describe how the sheep should sense its environment, how it should reason about sensed data and how it should act. This tutorial proved much harder than the first tutorial, but some students succeeded in implementing simple following behaviours to create 'pet sheep'.

The final project in this course was a collaborative task titled *You Versus the World*. Students were asked to implement a game with no humanoid or animal characters. Rather the primary antagonist was the environment itself. This project permitted students to explore the idea of interactive architecture as well as an alternative kind of place.

Six groups of students worked on this project. Some of their results are shown in Figure 4. Figure 4(a) shows the game *Clockwork* in which players must navigate their way through the internal workings of a clock; Figure 4(b) shows *Ziggurat* in which players must negotiate their way free of a series of traps inside a ziggurat. In another project, *The Heartless Galleon*, players must escape a pirate ship; and in the *Climate Game Challenge* players learn about and can interact with simple simulations of environmental phenomena.

**Fig. 4.** (a) Clockwork: players navigating through a clock; (b) Ziggurat: players escaping the traps inside a ziggurat.

The variety of designs produced reflects the enormous potential for virtual places to become dynamic, imaginative, interactive environments. These environments do not merely simulate the real world, but represent examples of virtual architecture that can itself interact with and respond to its inhabitants. These concepts have the potential to influence future real-world architecture by inspiring intelligent physical environments that can proactively support human activities.

### 3.3 Lessons Learned

The key issue associated with teaching interaction design is the combination of skills required by students, which include 3D modelling, design, computer programming and artificial intelligence. This could be overcome to some extent in this course, which included students with both design computing and digital media backgrounds. Students were surveyed at the beginning of the course and organised into cross-disciplinary project groups so they could utilise each other's skill sets in the projects. Students with digital media backgrounds would develop the 3D models while students with design computing backgrounds would focus on programming the behaviour of the designed artefacts. Communication and collaboration between students was critical for producing interactive worlds.

## 4 Future Directions for Designing and Learning in Virtual Worlds

Recognising and formalising the role of virtual worlds as novel design environments in their own right is the first step towards teaching the design of virtual worlds as a design subject. However, a number of other factors will also influence the development of such courses:

**Relevance to traditional design disciplines:** The unique capacity of virtual worlds to support the virtual organisation in industries such as commerce, health, defence, education and others makes them an important new technology and environment. There is thus a need for courses that teach the design of virtual worlds to move beyond the traditional gaming and social applications of virtual worlds to emphasise the broader role they can play in industry. This should include projects and exercises that require students to consider a broader range of applications.

**A need for designers with new skill sets:** The novel combination of skills required to achieve virtual place and interaction design calls, not only for new courses that teach virtual world design, but for new degrees to provide designers with skill sets that include design, digital media, computer programming and artificial intelligence. This combination of artistic and technical skills is particularly challenging to achieve because individual interests frequently lean in one direction or the other. Alternatively, courses that teach collaboration and communication in virtual worlds between designers with different backgrounds will be necessary. Such courses and degree streams are starting to emerge, but are not yet widely available.

**Improvements in virtual world technologies:** Finally, the technological development of virtual worlds will ultimately affect the utility of virtual environments. New virtual world technologies such as *Second Life*, *There* (There, 2009) and the recently trialled *GoogleLively* (Google, 2009) differ hugely in their capacity to support both place and interaction design. There are no current persistent virtual world technologies that genuinely combine the power of industrial strength 3D modelling software with industrial strength programming languages. Until such worlds emerge the full potential of virtual worlds is yet to be realised.

## Acknowledgements

Special thanks to students enrolled in the courses described in this Chapter for allowing the authors to use images of their designs.

## References

Baille-De Byl, P.: Programming believable characters for computer games. Charles River Media, Hingham (2004)

Champion, E., Dave, B.: Where is this place? Association For Computer Aided Design. In: Architecture 2002 Annual Conference, Pomona, USA, pp. 24–27 (2002)

Duggan, M.: The official guide to 3D game studio. Thomson Course Technology, Boston (2007)

Google: Lively by Google (2009), http://www.lively.com (Accessed January, 2009)

Kalay, Y.E.: Architecture's new media: Principles, theories, and methods of Computer-aided Design. MIT Press, MA (2004)

Kvan, T., Mark, E., Oxman, E., Martens, B.: Ditching the dinosaur: Redefining the role of digital media in education. International Journal of Design Computing 7 (2004)

Linden: Second Life (2009), http://www.secondlife.com Accessed January, 2009).

Meigs, T.: Ultimate game design: Building game worlds. McGraw-Hill/Osborne, Emeryville, CA (2003)

Maher, M.L., Gero, J.S.: Agent models of 3D virtual worlds, Association For Computer Aided Design. In: Architecture 2002 Annual Conference, Pomona, USA, pp. 127–138 (2002)

Maher, M.L., Merrick, K.: Agent models for dynamic 3D virtual worlds. In: The 2005 International Conference on Cyberworlds, Singapore, pp. 27–34 (2005)

Narayan, M.: Collaboration and cooperation: Quantifying the benefits of immersion for collaboration in virtual environments. In: VRST 2005. ACM Press, New York (2005)

Russel, S.J., Norvig, P.: Artificial intelligence: A modern approach. Prentice-Hall, Englewood Cliffs (1995)

There (2009), http://www.there.com (Accessed February, 2009)

Wooldridge, M.: Reasoning about rational agents. MIT Press, MA (2000)

# Sketch That Scene for Me and Meet Me in Cyberspace

Ellen Yi-Luen Do

Georgia Institute of Technology, U.S.A.

**Abstract.** This chapter discusses several interesting projects using sketching as an interface to create or interact in the 3D virtual environments.

**Keywords:** sketching, annotation, design, virtual environments.

## 1 Pick Up Your Purple Crayon

One evening, Harold decided to go for a walk in the moonlight, but there was no moon, so he drew one. He needed somewhere to walk on, so he drew a path. Anything Harold drew with his purple crayon became a reality. His dramatic adventures went as far as his imagination could take him to. There were beautiful landscape and scenery. Moving devices and living creatures were also abundant. He drew a boat when he found himself in trembling water, and drew a hot balloon to take him when he fell off from a mountain. He drew a fierce dragon to guard his apple tree, and friendly animals to eat the pies he couldn't finish on his picnic blanket.

In this beloved children's book (Johnson, 1955), Harold has the power to create a world of fantasy with his purple crayon. What would it be like if we all could pick up a purple crayon and create a world of our own by simply drawing it? This is no science fiction. Armed with creativity and computing technology, we could really design and interact with our own virtual environments by freehand sketching. This chapter will discuss the various roles sketching can play in the creation and use of virtual environments in design. In the following, Section 2 introduces research work on creating 3D virtual environment by freehand sketching. Section 3 shows the applications of sketching in 3D virtual environment as an interface to knowledge-based design systems. Section 4 presents case studies using virtual environments in collaborative design sessions. Section 5 concludes with discussions and future research directions.

## 2 Sketch That Scene for Me

If you can imagine it, you can build it. Well, that's true, but it's easier said than done. As more virtual environments are used for a wide variety of contexts such as learning, social networking and business, the ability to quickly and easily create three-dimensional virtual environments becomes increasingly important (Kessler, et al., 2000). Usually building a virtual world is a non-trivial task. One

X. Wang & J.J.-H. Tsai (Eds.): Collaborative Design in Virtual Environments, ISCA 48, pp. 121–130.
springerlink.com 

can either use traditional CAD modeling software and then covert the models to the appropriate format, or use the graphic editor supplied by the particular game, or directly code the scenes by hand (VRML, 1997; X3D, 2004; Java3D, 2005). All these methods have steep learning curves. They are usually time consuming and cumbersome and not easy to use (Chittaro, 2007). Wouldn't it be nice if we can create 3D content by simply sketching what we imagine?

## 2.1 Virtual Reality Sketchpad

Virtual Reality (VR) Sketchpad is a pen-based interface for rapid prototyping of 3D virtual environments (Do, 2001). This system provides an interface (in the form of a transparent window, or trace layers of a drawing board) for designers to draw diagrams to produce instant 3D worlds. The program recognizes simple geometric shapes such as circles and lines, and their spatial configurations as symbols for architectural objects such as walls, columns, a dinning table set with table and chairs, a lamp, a couch or a TV set. It then translates the hand-drawn input into 3D models arranged in corresponding locations according to the floor plan layout. The left image in Figure 1 shows the input on the drawing board window that is translated and converted into 3D virtual environment in a browser window on the right. The arrows on the floor plan indicated the places of interest and the specific viewing directions toward the scenes that are embedded into the browser and therefore define a viewing path into the 3D world for a guided tour. VR Sketchpad provides a quick and easy way for designers to see their 3D scene creations in a virtual environment that they can explore. Meanwhile, if the sensing option is activated, when the viewer navigates through the virtual environment, the embedded touch sensor and proximity sensor in the world's geometry on the client side could communicate with a Java applet running in the browser to continually report the browser's view location and orientation back to the sketching front end (Do, 2000) and thus provide feedback about usage pattern of the virtual environment. VR Sketchpad is a rapid prototyping tool for creating 3D virtual scenes. It directly interprets the drawing marks and delivers visualization (what you draw is what you get) for designers to explore and experience.

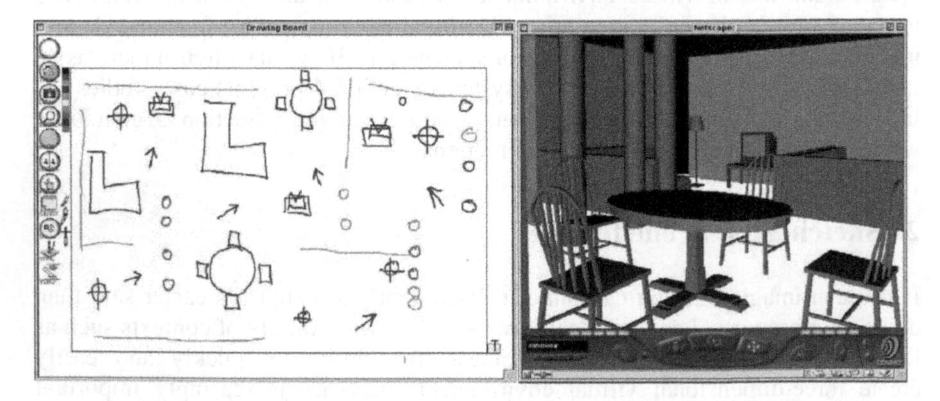

**Fig. 1.** A floor plan layout sketch creates a 3D world with architectural objects

### 2.2 Space Maker

"Architecture is the thoughtful making of spaces" (Kahn, 1957). Space is the essence of architecture. The art of architectural design is about arranging the void spaces, defined by solid architectural elements such as walls and floors, for people to inhabit (Berlage, 1908; Zevi, 1957). SpaceMaker is a symbol-based modeling tool that identifies spatial configurations in a bubble diagram to construct 3D virtual environment (Lee, 2003). Unlike VR Sketchpad that directly translates 2D drawn symbols into 3D objects for the interior space, SpaceMaker encourages designers to think about spaces first, to decide the essential characteristics for each space (i.e., enclosure level and privacy concerns), before constructing the 3D representation. A designer can assign the enclosure level (partition wall or columns) for each room, and a preference priority between rooms. For example, Figure 2 (top) shows a diagram with space bubbles (left) is translated into boundary objects with partition walls in a 2D floor plan (middle) and 3D world (right). As each functional space has a different privacy concern, a boundary conflict resolver checks the priority preference of adjacent rooms to assign the boundary enclosure condition (open or closed) with a colonnade or a partition wall (Figure 2 bottom). SpaceMaker provides more than just a simple visualization tool. It helps designers to think about model detailing, and helps them "make spaces."

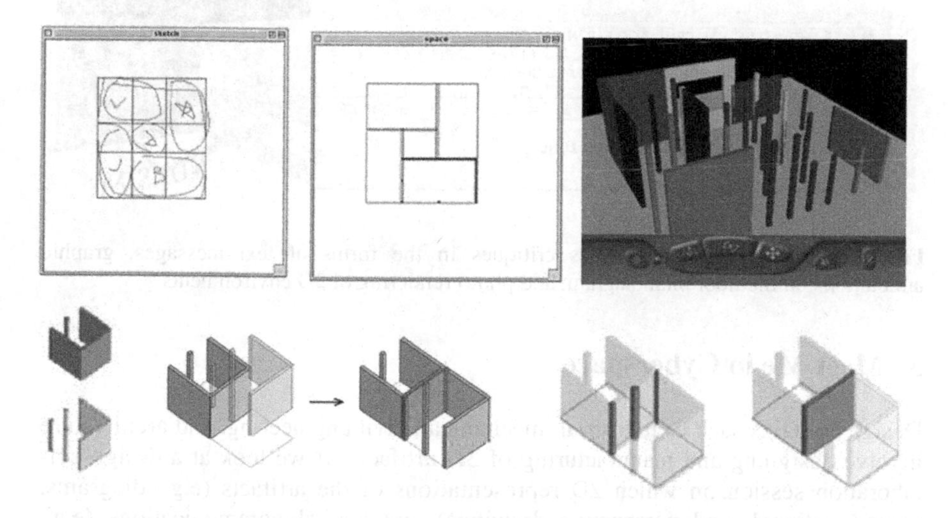

**Fig. 2.** SpaceMaker translates bubbles into rooms with walls, and a 3D virtual environment. The enclosure level (open or closed, with columns or walls) between two rooms is resolved by checking the room privacy preference priority rule.

### 2.3 Design Evaluator

Design is a reflective practice. The "reflection-in-action" design process is a "see-move-see cycle" that involves sketching, evaluating, and modifying of design ideas (Schön, 1985; Schön and Wiggins, 1992). Designers draw to externalize their thinking (Tversky, 1999) and to have a conversation with their ideas (Laseau, 1980; Goldschmidt, 1991). If drawings could reason and talk back, what critiques will they give? Can we have critiquing sketches to help us reflect and make better design decisions? Design Evaluator incorporates critiquing into a freehand sketch design system (Oh, 2004). The system interprets the floor plan diagrams and checks against built-in spatial rules to provide critiques in the forms of text message, annotated drawing and 3D model/walk-through. Figure 3 shows a hospital layout sketch receives critiquing about the circulation path between ICU and ER with a graphical annotation of the path (top left), a text message (directly below), a 3D visualization (top right), and photo-realistic image mapping (bottom right). The drawing also receives a zoning suggestion (top middle and bottom left)

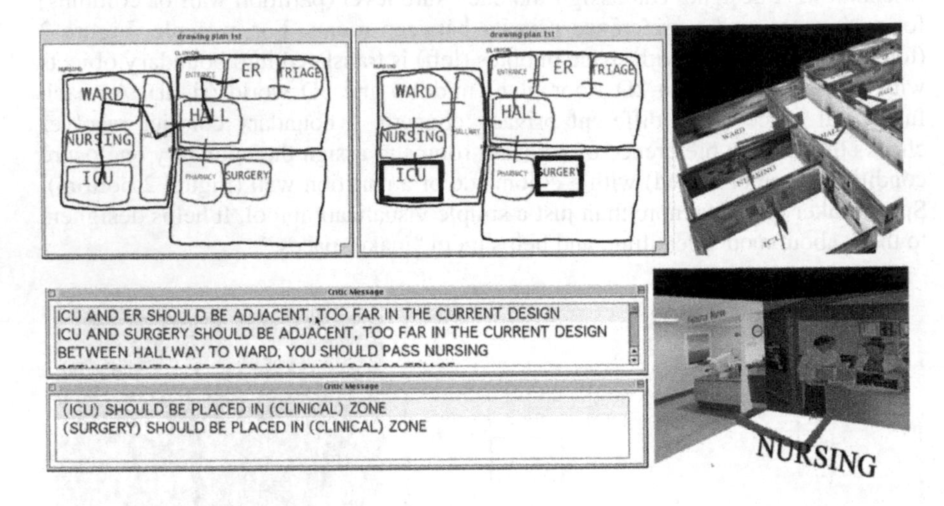


**Fig. 3.** Design Evaluator provides critiques in the forms of text messages, graphic annotations on the floor plan diagram, and photo rendering of 3D environments

## 3 Meet Me in Cyberspace

Design domains such as industrial, mechanical, civil engineering, and architecture involve designing and manufacturing of 3D artifacts. If we look at a design collaboration session, in which 2D representations of the artifacts (e.g., diagrams, plans, sectional, and perspective drawings) and textual communications (e.g., phone, fax, instant messaging, email, etc) are used, we can often find comments like “That's not what I meant!” or “This is in the wrong place!” These problems arise because we can’t see the views of our collaborators or the exact locations they are pointing at. A 3D annotation system can easily address the “wish you were here” problem. This system can also play the role of a collaborating partner, a helpful assistant or an expert advisor in a design process. Wouldn’t it be nice if we could use sketching as an interface to access knowledge-based design tools in a virtual world?

### 3.1 Space Pen

Space Pen provides annotation capabilities in a 3D virtual environment for asynchronous online design collaboration (Jung, 2001; Jung et al., 2002). The Space Pen server converts any design model uploaded into a Java 3D model for viewing in a standard Web browser. It also provides a platform for drawing onto and into 3D models so that collaborating design team members can browse and annotate on the model with graffiti-style sketching and Post-It® style tags. Designers can attach text annotation notes to any surface of the 3D model and starts a "location-based" threaded discussion with links to issues and authors. The Space Pen server automatically sends emails to inform all related stakeholders when a new annotation note is made "on the spot." Space Pen also has simple stroke recognition to identify figures such as arrows, circles, and rectangles. Recognized sketch objects can be rectified as model geometry or interpreted as commands. Furthermore, designers can mark on any existing model surfaces or on a temporary drawing plane to add geometry to the model. For example, Figure 4 (left) shows a new wall extension sketched on the temporary wall. Figure 4 (right) shows a designer reviewing an architectural design "takes out the red pencil" and draws on the wall to suggest a new location for the window.

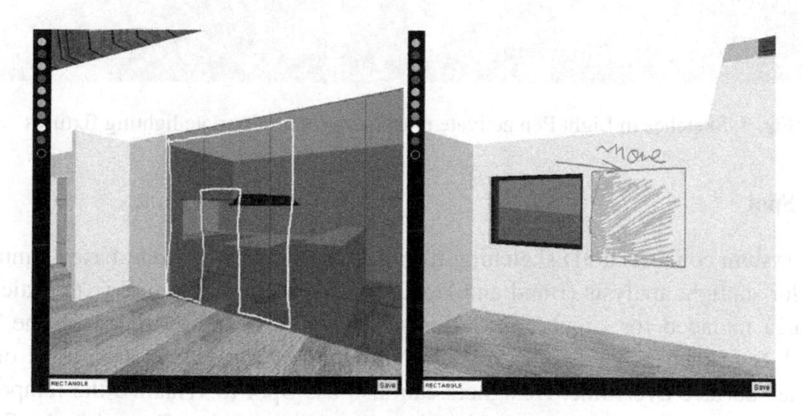


**Fig. 4.** Space Pen supports sketch annotation on temporary surface invoked by a gesture command (left) or on an existing surface in the 3D model (right)

### 3.2 Light Pen

If walls could talk, what design help would they offer? Can lighting designers paint with light in a room and have the walls figure out where to put lamps to get the effect they want? Light Pen lets designers sketch with light (Jung, 2003). It is a lighting design system driven by sketching on 3D virtual models. The Light Pen system uses sketching in 3D as input to a knowledge-based lighting design decision-support system. This is similar to the idea of using sketching and diagramming to interact with knowledge-based design systems in the form of posing queries to visual databases or setting scenarios for simulations (Gross and Do, 2000; Do, 2005). However, Light Pen extends this work to sketching on a three dimensional model. The designer specifies where illumination is desired by sketching directly on surfaces in a 3D model, and Light Pen selects and places the light sources based on a rule-based electrical lighting fixture advisor and then visually renders their effects. Figure 5 shows that based on the desired lighting and the model geometry Light Pen infers the contexts, recommends solutions, and then selects fixtures from a catalog based on their desired characteristics and adds the fixtures to the 3D model to indicate proposed design solution.

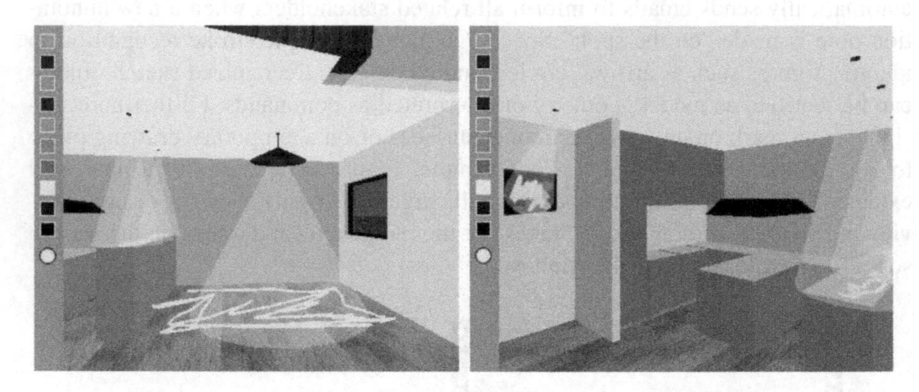

**Fig. 5.** Sketches in Light Pen activate placements of appropriate lighting fixtures

### 3.3 Spot

Spot system connects a 3D sketching front end on the web to a rule-based simulation for sunlight analysis (Bund and Do, 2005). As shown in Figure 6, to indicate the area intended for simulation designer sketches a boundary shape on the 3D model. Spot then generates the spatial distribution of the illumination level on a selected surface over time. Designers can also use Spot to visualize the temporal information of light distribution over time for a given point. For example, Spot generates a calendar diagram of a chart where the X and Y axes represent the months of the year and the time of the day for each point tapped on the 3D model. The color of each calendar cell is the result of the calculation of the light amount reaching the specific point. Spot generates time projection and navigable animation. Designer can adjust the spatial variables (x, y, z) of 3D geometry with standard interface (mouse, arrow keys or joystick), text annotation and sketching (pen and tablet). The temporal variables (date and time) are displayed in additional views with the appearance of a graphic calendar. The resulting sunlight simulation (in gradient distribution) is displayed on the 3D environment (Figure 6 right).

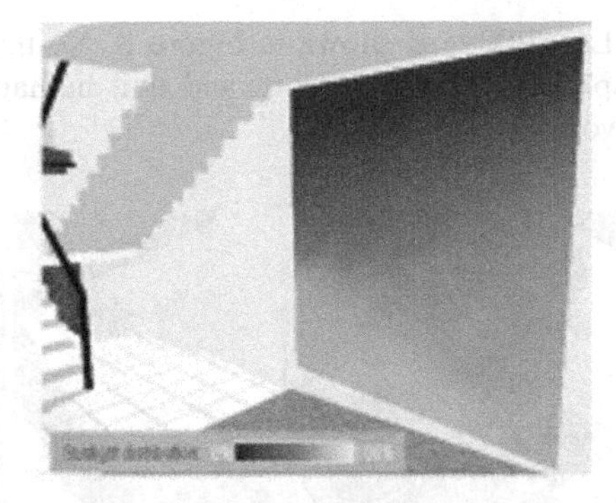

**Fig. 6.** A boundary area sketch (left) on the wall activates sunlight distribution simulation (right)

# 4 Working Together While Apart

Virtual Environments are good platform for communication and information sharing between distributed design teams (Simoff and Maher, 2000; Hinds and Bailey, 2003). For asynchronous collaboration, annotations of design intentions and alternative design options can be provided in virtual environments for different stakeholders to review and discuss. For example, Figure 7 shows the Redliner project used to support two real design cases.

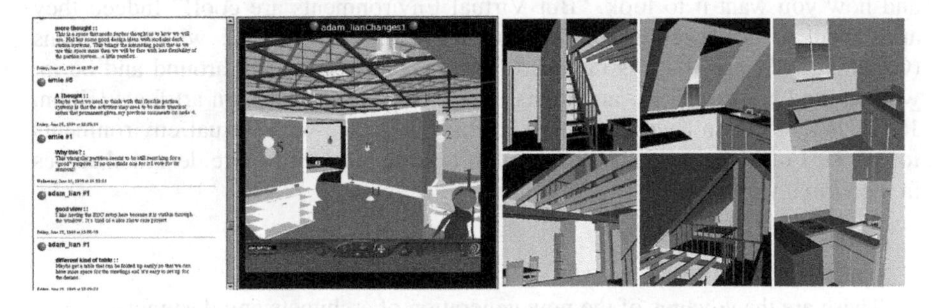

**Fig. 7.** Left: Redliner showing comments and annotation markers in the space. Right: Providing different design alternatives and viewpoints in apartment remodeling.

Figure 7 left shows a computer laboratory redesign with comments from the residents and designers in Redliner. Figure 7 right shows the renovation of an apartment building in which different spatial treatments (e.g., options of skylight, attic, and window location, etc) were proposed and posted on the Redliner for the owner, contractor, and design partners for decision-making.

It is also noted that even though different virtual environments (e.g., Unreal, Second Life) may support synchronous collaboration of remote team members meeting online, sometimes people may result to "low tech" solution such as pointing the webcams at the screens to share their 2D drawing or 3D graphic models

(Lee, 2009) as shown in Figure 8. No matter how cool tools are, there is always space for improvements, and that human can always find ways to make things work!

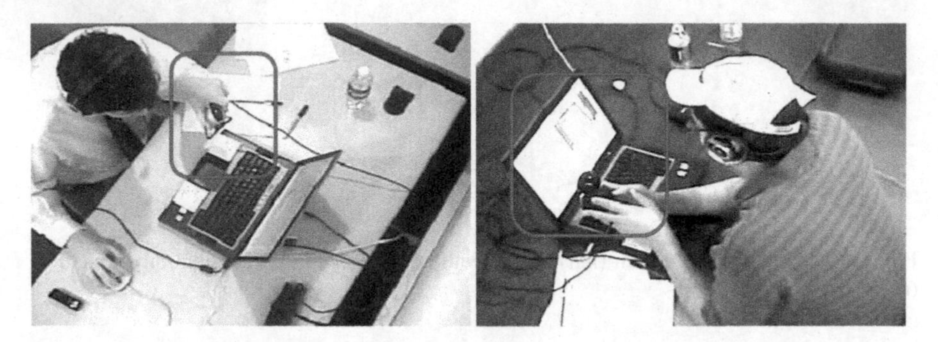

**Fig. 8.** Collaborating team members put the webcams in front of their screens when discussing with their collaborators

## 5 Space: The Final Frontier

You may find it odd that we are discussing sketching in the context of collaborative design in virtual environments. Why not? Paper and pencil is the quintessential direct manipulation interface—you draw what you want, where you want it, and how you want it to look. "But Virtual Environments are cool!" Indeed they are. There is the necessity of drawing for architects to think with their pens (Graves, 1997), and there is the need for an art you can "walk around and be in, walk around and be in" for architects to contemplate their design artifacts (Kahn, 1972). Why shouldn't we integrate sketching interface with virtual environments for the best of both worlds? Why shouldn't we engage with more design sketches in the virtual worlds?

Let us begin with a story:
Space: The final frontier
These are the voyages of the new generation of architects and designers
Their lifelong mission
To explore strange new worlds
To seek out new possibilities and new challenges
To boldly go where no one has gone before

## References

Berlage, H.P.: Grundlagen und Entwicklung der Architektur (Principles and Evolution of Architecture), Rotterdam/Berlin, p. 46 (1909)

Bund, S., Do, E.Y.L.: SPOT! Fetch Light: Interactive navigable 3D visualization of direct sunlight. Automation in Construction 14(2), 181–188 (2005)

Chittaro, L., Ranon, R.: Web3D technologies in learning, education, and training: Motivations, issues, and opportunities. Computers and Education 49(1), 3–18 (2007)

Do, E.Y.L.: The Right Tool at Right Time: Investigation of Freehand Drawing as an Interface to Knowledge based Design Tools. PhD, Georgia Institute of Technology (1998)

Do, E.Y.L.: Sketch that scene for me: Creating virtual worlds by freehand drawing. In: eCAADe 2000, pp. 265–268 (2000)

Do, E.Y.L.: VR sketchpad: Create instant 3D worlds by sketching on a transparent window. In: de Vries, B., van Leeuwe, J.P., Achten, H.H. (eds.) CAAD Futures 2001, pp. 161–172. Kluwer, Dordrecht (2001)

Do, E.Y.L.: Design sketches and sketch design tools. Journal of Knowledge Based Systems 18, 383–405 (2005)

Goldschmidt, G.: The dialectics of sketching. Creativity Research Journal 4(2), 123–143 (1991)

Gross, M.D., Do, E.Y.L.: Ambiguous intentions: A paper-like interface for creative design. In: ACM Conference on User Interface Software Technology, pp. 183–192 (1996)

Gross, M.D., Do, E.Y.-L.: Drawing on the back of an envelope: a framework for interacting with application programs by freehand drawing. Computers & Graphics 24(6), 835–849 (2000)

Graves, M.: The necessity of drawing: Tangible speculation. Architectural Design, 384–394 (June 1977)

Hinds, P.J., Bailey, D.E.: Out of sight, out of sync: Understanding conflict in distributed teams. Organization Science 14, 615–632 (2003)

Java 3D: The Java3D API (2005), http://java.sun.com/products/java-media/3D

Johnson, C.: Harold and the purple crayon. Harper Collins Publishers, New York (1955)

Jung, T., Gross, M.D., Do, E.Y.L.: Space pen: Annotation and sketching on 3D models on the internet. In: CAAD Futures 2001, pp. 257–270 (2001)

Jung, T., Gross, M.D., Do, E.Y.L.: Annotating and sketching on 3D web models. In: Proceedings of the IUI 2002 Conference, San Francisco, January 13-16, pp. 95–102 (2002)

Kahn, L.: Architecture is the thoughtful making of spaces. Perspecta IV, 2–3 (1957)

Kahn, L.: How'm I doing, Corbusiner? The Pennsulvania Gazette 71(3), 19–26 (1972); reprinted in Latour, A. (ed.), Louis I. Kahn – Writings, Lectures, Interviews, Rizzoli, pp. 297-312

Kessler, G.D., Bowman, D.A., Hodges, L.F.: The simple virtual environment library: An extensible framework for building VE applications. Presence 9(2), 187–208 (2000)

Laseau, S.: Graphic Thinking for Architects and Designers. Van Nostrand Reinhold, New York (1980)

Lee, M.L., Do, E.Y.L.: SpaceMaker - Creating space by sketching it. In: ACADIA 2003, pp. 311–323 (2003)

Lee, S., Ezer, N., Sanford, J., Do, E.Y.L.: Designing together while apart: The role of computer-mediated communication and collaborative virtual environments on design collaboration. In: IEEE SMC 2009, Systems, Man and Cybernetics, pp. 3593–3598 (2009)

Oh, Y., Gross, M.D., Do, E.Y.L.: Critiquing freehand sketches: A computational tool for design evaluation. In: Gero, J.S., Knight, T. (eds.) Visual and Spatial Reasoning in Design III [VR 2004], pp. 105–120 (2004)

Schön, D.: The Design Studio. RIBA Publications, London (1985)

Schön, D., Wiggins, G.: Kinds of seeing and their function in designing. Design Studies 13(2), 135–156 (1992)

Simoff, S.J., Maher, M.L.: Analysing participation in collaborative design environments. Design Studies 21(2), 119–144 (2000)

Tversky, B.: What does drawing reveal about thinking? In: Gero, J.S., Tversky, B. (eds.) Visual and Spatial Reasoning in Design, Sydney, Australia: Key Centre of Design Computing and Cognition, pp. 93–101 (1999)

VRML International Standard: VRML 1997 Functional specification and VRML 1997 External Authoring Interface (EAI) International Standard ISO/IEC 14772-1:1997 and ISO/IEC 14772-2:2002 ISO/IEC 14772-1:1997 and ISO/IEC 14772-2:2002 (1997), Available from http://www.web3d.org/x3d/specifications/vrml/

X3D International Standard: X3D framework & SAI. ISO/IEC FDIS (Final Draft International Standard) 19775:200x (2004), http://www.web3d.org/x3d/

Zevi, B.: Architecture as Space: How to Look at Architecture. Horizon Press, New York (1957)

# A Hybrid Direct Visual Editing Method for Architectural Massing Study in Virtual Environments

Jian Chen

Brown University, Providence, USA

**Abstract.** This chapter presents a hybrid environment to investigate the use of a table-prop and physics-based manipulation, for quick and rough object creation and manipulation in three-dimensional (3D) virtual environments (VEs). A set of new direct visual editing techniques were designed to model virtual objects. The system has been integrated into a Cave Automatic Virtual Environment (CAVE) and a large screen display called GeoWall to address early architectural design called massing study. Experimental results demonstrate the following findings: (1) the physical prop for the CAVE and the GeoWall is an effective way to interact with VEs, at least for the tasks that have been studied; (2) architects can quickly model virtual building masses using our techniques; and (3) physics need to be combined with constraints in order to be effective.

**Keywords:** hybrid environment, computer-aided massing study, physics-based object manipulation, interactive 3D immersive environment, direct visual editing.

## 1 Introduction

Achieving content creation and modification is a major goal of immersive design and has its potential to have broad impact on architecture, mechanical engineering, and automobile industries. One of the content creation tasks is called massing study. A scenario of use is similar to the following: compose a building according to a particular shape and style and then construct a mass. The conventional practice is to cut cardboard or shape clays to construct physical miniature mock-ups. Unlike other modeling systems, the purpose is not to draw fine delicate strokes but to examine shapes and forms.

One obstacle remains to accomplish massing tasks in immersive virtual environments (VEs). There is a lack of easy-to-use rough and quick content creation. Several 3D immersive design tools (see Deisinger et al., 2000 for a summary) allow for direct manipulation on shapes and forms. However, for the environments that are represented with only one scale, directly editing 3D scenes could introduce a performance penalty due to transitional costs between the large view

X. Wang & J.J.-H. Tsai (Eds.): Collaborative Design in Virtual Environments, ISCA 48, pp. 131–140.
springerlink.com

(for editing) and the overview (for checking forms and shapes). Such transitional costs for interaction may cause fatigue associated with spatial input (Hinckley et al., 1994).

Advances in modeling and rendering notwithstanding, designers continue to favor their conventional studio style for massing study. A VE appeals as an artistic medium to show objects at multiple scales. What is needed is a seamless transition from architects' office to immersive design environments. Such a transition should be further augmented rather than hindered by non-conventional input devices to make massing tasks more efficient and effective, as compared to the physical one. This is precisely our focus in this chapter.

The challenge for the design of the 3D massing study system lies in interaction. An issue is to find an appropriate mapping from high degree-of-freedom (DOF) input devices to high DOF modeling tasks that would minimize user's attention on user interfaces. Though direct manipulation within a VE on architectural modeling remains by far the dominant interaction paradigm, we propose a hybrid environment and gesture-based direct visual editing method (see Figure 1) (Chen, 2006). Rather than operating on widget, direct visual editing merges command input and actions on objects to eliminate the extra level of widget input abstraction. The transitions between actions are smooth and the UI (user interface) supports close-body interaction in a relatively small working volume. Our system automatically interprets a command according to the handedness, number of objects grabbed, and motion of the user's gesture. Using our techniques, the user can scale, move, rotate, quickly stack, align, delete, and retrieve architectural elements. Both a miniature view and a large one-to-one scale scene are presented to reduce the transitional cost between views.

This chapter presents the design and results from a user study, the present study of using a hybrid system and 3D direct visual editing can serve as a useful guide and starting point for the community of designers and practitioners who wish to investigate rich design space for modeling.

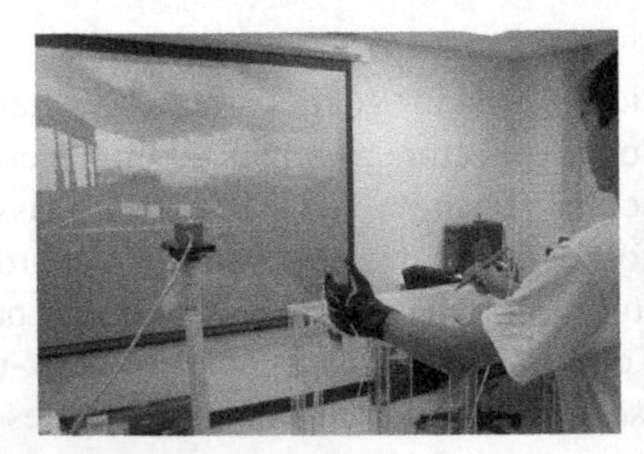

**Fig. 1.** A hybrid massing study environment coupled with a large screen display (GeoWall, left). In such an environment, the two hands function differently. The non-dominant hand wears a pinch glove for imprecise and quick input and the dominant hand holds a stylus pen to conduct relatively precise object manipulation. The hybrid system also works in the CAVE with a miniature view displayed on the tabletop (the right image).

## 2 Related Work

Content creation is one of the most important challenges in computer graphics and immersive modeling. Most immersive conceptual design systems (Bowman, 1996; Deisinger et al., 2000) use a menu for primitive creation, alignment, selection, and reshaping and operate on a large design space. Our design removes the menu selection step and instead uses 3D gesture input because merging command and object selection could improve task performance (Guimbretière et al., 2005).

Tangible user interfaces (TUI) simulate architectural desktop environments to allow architects to draw directly on paper using digital or regular ink or using sensors to push and pull a model (Lee et al., 2006). This provides excellent haptic feedback. One difficulty of using such a system for massing study, however, is to undo or delete objects. Other immersive design systems make use of domain specific characteristics to design interaction for immersive tasks (Underkoffler and Ishii, 1999; Chen et al., 2004). We share the same goal in capturing domain characteristics in the design process, however, we design interaction for the massing tasks that impose different design requirements, which have not been sufficiently studied.

There are techniques and tools that support conceptual design by editing two-dimensional (2D) inputs to construct three-dimensional (3D) structures. One method is to use free handing sketching recognition (Zeleznik et al., 1996; Schkolne et al., 2001). This method has low overhead in representing, exploring, and communicating geometric ideas. It is analogous to using physical pencil and paper, which are probably the best tools to illustrate design ideas, despite that recovering 3D shapes from 2D drawing is challenging (Chen et al., 2008). Teddy and its extensions allow users to draw curves and support many other operations, such as extrusion, cut, erosion, and bending for modification purpose (Igarashi et al., 1999), though the systems are limited to making blob style objects.

An alternative modeling method is to sketch directly in 3D. The main approach is to operate on 2D shapes to derive the 3D ones using constraints (SketchUp, 2008). For example, SESAME advances the design by minimizing the 2D input mode and facilitates efficient drawing with suggestions, automatic segmentation, and recognition of closed structures (Oh et al., 2006). Our work tries to achieve the same goal but emphasizes direct 3D editing, so that designers do not have to translate 2D drawings into 3D in their minds. In addition, our system is designed for immersive displays, which could incur the change of gesture grammar.

## 3 Designing for Massing Study

In order to design an effective massing study environment, we started with analyzing prior work and consulting architects to learn the design requirements. We interviewed professionals in architectural firms and people from the architectural department on campus to elicit design requirement. We asked questions about current practices and conventional workflows, and pros and cons of existing modeling tools. Finally, we obtained important design elements in the massing study process and identified a set of principles for interactive content creation.

### 3.1 Rough and Quick Content Creation

The purpose of massing study is not to draw fine delicate strokes to indicate detailed structures, but rather to apply a series of well-established architectural elements for making mock-ups. Architects make frequent remodelling, reconstruction, design comparison, and the addition and removal of clay or cardboard masses. Box, sphere, pyramid, stairs, and bars are basic shape elements (Hohauser and Demchyshyn, 1984). Objects in the scene should be solid models.

### 3.2 Visual Context and Multiple Scale Viewing

When creating objects, architects like to play with shapes and their spatial layout. They then place the finished design in the surrounding context to examine the form. Architects prefer to examine their design in a one-to-one scale. In our design, we present both a miniature and a one-to-one scale view. The miniature view shows the objects being edited. The real scale view includes alternative designs and surroundings. We choose to have two views also out of design considerations for 3D UIs. Operating in a close-body fashion can also reduce fatigues (Hinckley et al., 1994) and increase input precision (Zhai et al., 1996).

### 3.3 Ergonomic Requirements

One major drawback of current VE design is the ergonomic issue. Users will not use a system that requires standing posture, lacks arm-resting positions, and exploits precision fine motor skills.

## 4 System Overview

To address users' requirements, we built a hybrid environment by integrating a table prop to the conventional immersive or semi-immersive VEs. A table prop is integrated in a 10'x10'x10' CAVE or in front of a large screen display, called Geowall (Consortium). The table is transparent made out of acrylic glasses and thus does not block the user's view. It is tracked with a Polhemus FastTrak, therefore becomes an active interface component. While coupled with the CAVE, a virtual table was rendered at the same location as the physical table. Widgets and miniatures of the virtual scene can therefore be displayed on the physical table. The peripheral vision provided by CAVE or a large display presents the context of the design in a one-to-one scale.

The idea of using the table-prop was driven by two considerations: one for the architectural domain of use and the other for the interactivity. The table utilizes architects' familiarity with drafting on their workbench and naturally divides the design space into two areas, one on the table (a miniature view) and the other in the immersive VE (a one-to-one scale).

The table prop could support precise object manipulation and close body interaction to avoid performance loss due to the lack of a physical surface to touch with free hand manipulation. Close-body two-handed manipulation techniques can improve task performance and precision of interaction by providing proprioceptive cues and reduce repetitive actions by operating on miniatures. The table also shares the benefits of passive haptic feedback similar to the pen-and-tablet metaphor user interface, and frees the user's non-dominant hand for other operations.

A tracked pinch glove and a stylus pen are integrated to support asymmetrical two-handed manipulation. The user will hold the pen in his or her dominant hand and wear the glove in his or her non-dominant hand. By doing so, users can perform both tasks requiring imprecise but quick input (e.g., grasping) using the glove and tasks requiring fine operations using the stylus. Our interaction design is intended to support direct visual editing with gesture input.

# 5 Direct Visual Editing Techniques

We define direct visual editing as the type of direct manipulation in which a user's action is applied to visual objects using gestures rather than widget input. We differentiate this method with other direct manipulation such as widget-based direct manipulation of visual content, in that direct visual editing merges command selection and object manipulations. Such a merger has shown its benefits in 2D menu manipulation because the interaction reduces the access cost of the menu system.

## 5.1 Object Creation

One way to create object is to use what is called copy-by-example (see Figure 2). Every object that has been placed in the scene can be copied. Our interaction technique directly maps the orientation of the user's hand to the system actions. The user points the index finger horizontally (+/- 45 degree error permitted) and then pinches index finger and thumb to create a copy. The finished copy is attached to the user's hand. This action is similar to pulling another object out of existing ones. Any objects, including grouped objects, can be created in this manner.

The advantage of copy-by-example is that it simplifies the architect's workflow. Rather than building a structure from scratch, the user can generate a copy of a structure and work from there to construct a similar one. This operation is useful to build alternative designs, when many similar parts exist.

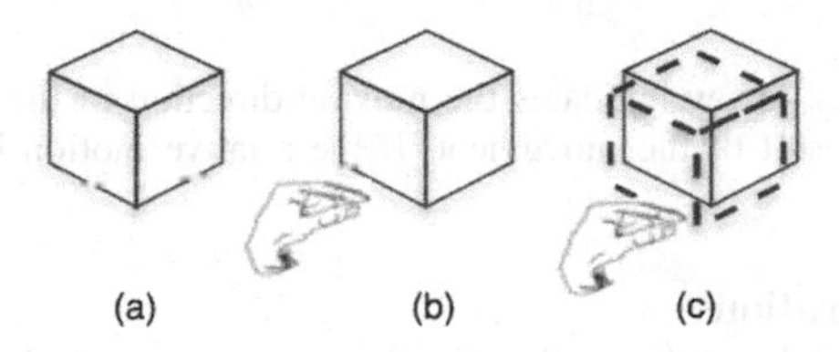


**Fig. 2.** Copy by example. (a) Original state. (b) The user moves his / her hand closer to the object and the index finger is near horizontal to the supporting plane. (3) The user pitches the index figure and thumb to create an object.

## 5.2 Object Manipulation

Another single-handed operation is manipulation (see Figure 3). The system uses the grasping gesture, i.e., downward direction from the output of the tracker attached to the pinch glove (+/- 45 degrees error permitted). To make the grasping easier, no "virtual touch" is required. A threshold of 10 cm distance can activate the "grab" action. An alternative way to move an object is to use the hand to grab and move. Users can also pass an object from one hand to the other.

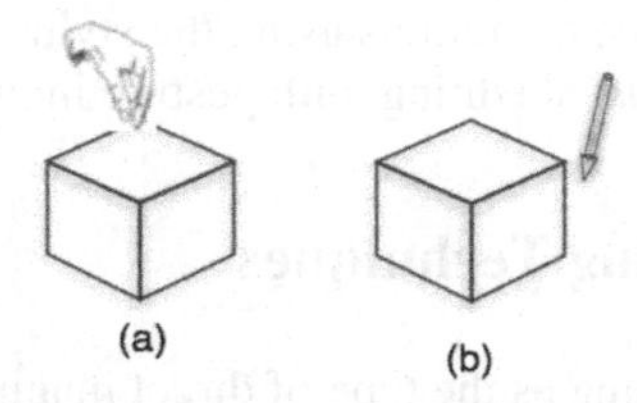


**Fig. 3.** Single-handed manipulation gesture. (a) Grasping using the tracked glove. No direct touch is needed. The system defines the action based on the direction of the hand. (b) Moving using the stylus.

### 5.2.1 Single Axis Scaling and Rotation

Scaling and rotation make use of two-handed manipulation (see Figure 4). The user places his or her two hands on each side of the selected object. Only one axis of rotation or scaling is enabled to reduce the degree-of-freedom of the spatial input. The center of the rotation and scaling is the origin of the local coordinate system. The axis of rotation is perpendicular to the rotational plane of the two hands. The system automatically chooses rotation or scaling based on the relative motion of the two hands. When the distance between two hands is larger than the rotational distance, the system will scale the object, otherwise, a rotation command is issued.

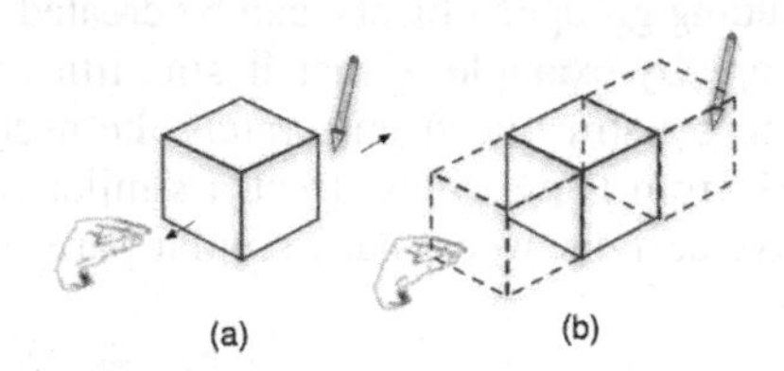


**Fig. 4.** Scale object. (a) Arrow indicates the moving direction of the user's hand. (b) The object is scaled as a result of the movement. If the relative motion is rotation in (a), the object will be rotated.

### 5.2.2 Boolean Operations

Similar to scaling and rotation, the Boolean operation utilizes two hands (se Figure 5). Three Boolean operations are supported: union, difference, and intersection. This command is activated when each hand pinches and grabs an object.

The object in the right hand will be the operand. Once two objects are grabbed, the system automatically changes to the Boolean state. Therefore, the user does not need to keep the index finger and thumb pinched in order for the object to remain grabbed.

The user pinches the index and the middle fingers to shuffle between Boolean operations. A preview shows the results of the Boolean operation at run time. Pinching ring finger and thumb will also enter the preview model and disable Boolean operation. Clicking the stylus button will complete the current operation. Boolean operation can also be performed in place using the objects grabbed in the user's right hand by pinching the index finger and thumb (with no object selected).

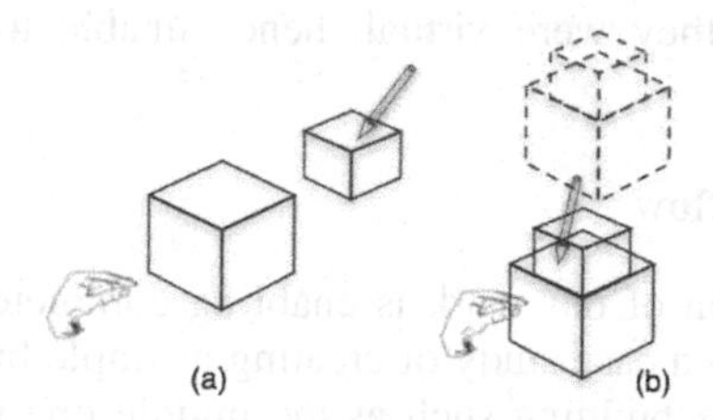


**Fig. 5.** Union and subtraction of any shapes. (a) Grabbing an object in each hand will activate the Boolean command. (b) Union, what you see is what you get. A preview is shown to draw the result.

### 5.2.3 Off-the-Table Deletion and Retrieval

Users can "push" the object off the table to delete it. The object will be placed on the "ground". This placement does not need to be precise because objects behave according to physics laws. The object will fall due to gravity until it hits another surface, such as the physical floor in the CAVE. To retrieve the object, users use a ray-casting technique to grab the object; the object is animated to the pen-tip and can be placed back onto the tabletop.

This technique can provide easy object composition, deletion, and retrieval by declaring the table-prop the current working place. All other places can be considered as an extra storage space. The action is similar to throwing garbage into a trashcan on the floor. Users can also use this method to save the table space by placing extra objects on the floor.

## 5.3 Physics

We implemented constrained physics-based manipulation. Three types of physical behaviors are supported: kinematics interaction, collision detection, and gravity. Kinematics interactions involve one object's motion that constrains or affects the position and orientation of another object at some connection point or joint. Collision detection prevents objects from penetrating each other while positioned. Objects having gravity allows causal placement of an object on top of another.

These physical behaviors can allow the architect to predict how objects move relative to each other. Physics-based manipulation can help rough placement of objects in space, as positioning in six (DOF) simultaneously is still difficult even with all the benefits from the constraints provided by the table-prop. For example, objects can float in space or penetrate each other, and making virtual objects touch and align is not easy.

The system automatically turns off the kinematics constraints when an object is grabbed and turns it back on when the object is released. Therefore, objects can penetrate each other during maneuvering in our current implementation. This design decision was made after a pilot study, where we found that full physics did not work well for object manipulation, because the existing structure could be toppled if a large object was picked up and accidentally hit the pre-built structure. While this "accidental" hit might not happen in the physical world, it happened quite often in the virtual world possibly because virtual objects did not provide tactile feedback and they were virtual, hence unable to convey their physical behaviors.

### 5.4 Modeling Workflow

One major contribution of our work is enabling complete visual editing of mass quickly. We will show a case study of creating a simple building from scratch. To create a more complex building such as the middle one in Figure 6, middle, the user would need to perform the following actions: (1) grabbing a virtual box from the knapsack and resizing to the desired dimensions; (2) grabbing another one with stylus pen and performing a Boolean operation to create the hole and place it on the table; (3) grabbing and scaling several more boxes to stack them together to form the building up front; (4) grouping the mass by clicking the lock button on the bottom right corner of the table; and (5) rotating the block to the desired orientation.

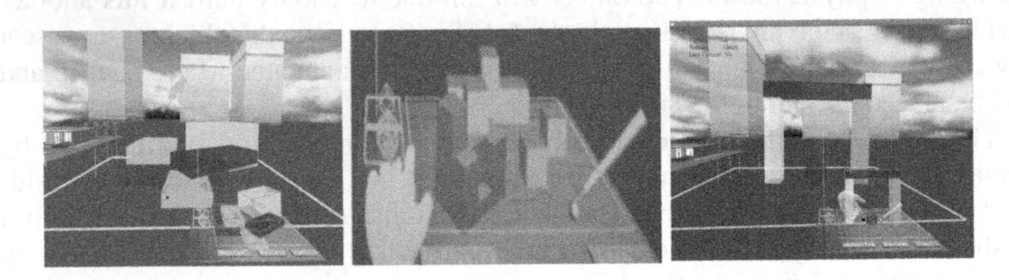

**Fig. 6.** Example scenes created using our system

## 6 Usability

We conducted an exploratory user study to examine the usability of our designed system and techniques. Constrained by the page limit, we only report major results. At a higher level, we wanted to learn overall subjective responses to the design of the table-prop, physics, and direct visual editing methods. All participants were positive about using the table and the seated condition. Major results include the following.

Firstly, the physical table warrants passive haptic feedback and close-body interaction that are well suited for the massing study, at least for the tasks we measured. When users acted at the table level, the table-prop helped stabilize their hands and served as an anchor surface for transformation or rotation. However, the table-prop did not help with the upper level construction when participants tended to move their hands much higher than the table.

Secondly, the table surface increases the stability of users' hands, reduces excessive effects of undesired input actions and enables fine movement at the finger level. 2D WIMP and 3D interaction were permitted.

Thirdly, physics-based direct visual editing provides effective scene composition. Full Newtonian physics might not be a good option for VEs. Instead, constrained physics work better. Physics hindered some user tasks. One user commented that architects sometimes build models from top to bottom. Physics-based UI would not allow him to do so because objects tend to fall onto the desk though it is consistent with what could happen in a design studio. One solution is to lay the space on top of the drafting table to define different levels of operation.

## 7 Conclusions

This chapter presents our experiences in the design and evaluation of a hybrid virtual environment system and direct visual editing techniques for architectural massing study. The chapter contributes (1) the design of a hybrid environment for massing study interaction and an exposition of the underlying design rationale, (2) advancement to direct visual editing techniques, and (3) design experiences and modeling method.

## Acknowledgement

The work was sponsored in part by grants from NSF-IIS-0237412 and NSF-DUE-0817106. The author is grateful to Prof. Doug A. Bowman for directing this research, to Professors Robert, P. Schubert, Marte Gutierrez, Mehdi Setareh, Walid Thabet, Heiner Schnoedt, Dennis Jones, and the participants for their input, support, time, and efforts.

## References

Bowman, D.A.: Conceptual design space - beyond walkthrough to immersive design. In: Bertol, D. (ed.) Designing Digital Space, pp. 225–236. Wiley, New York (1996)

Chen, J.: Design and Evaluation of Domain-specific Interaction Techniques in the AEC Domain for Immersive Virtual Environments, unpublished PhD dissertation, Department of Computer Science, Virginia Tech (2006)

Chen, J., Bowman, D.A., Lucas, J.F., Wingrave, C.A.: Interfaces for cloning in immersive virtual environments. In: Proceedings of Eurographics Symposium on Virtual Environments, pp. 91–98 (2004)

Chen, X., Kang, S., Xu, Y-q, Dorsey, J., Shum, H.-Y.: Sketching reality: realistic interpretation of architectural design. ACM Transaction on Graphics 27(2) (2008)

GeoWall Consortium, `http://www.geowall.org/` (accessed in May 2009)

Deisinger, J., Blach, R., Wesche, G., Breining, R., Simon, A.: Towards Immersive Modeling - Challenges and Recommendations: A Workshop Analyzing the Needs of Designers. In: Proceedings of Eurographics Workshop on Virtual Environments, pp. 145–156 (2000)

Guimbretière, F., Martin, A., Winograd, T.: Benefits of merging command selection and direct manipulation. ACM Transactions on Computer-Human Interaction (TOCHI) 12(3), 460–476 (2005)

Hinckley, K., Pausch, R., Goble, J., Kassell, N.: A survey of design issues in spatial input. In: Proceedings of ACM Symposium on User Interface Software and Technology, pp. 213–222 (1994)

Hohauser, S.: Architectural and interior models, Van Nostrand Reinhold (1984)

Igarashi, T., Matsuoka, S., Tanaka, H.: Teddy: a sketching interface for 3D freeform design. In: Proceedings of the 26th annual conference on Computer graphics and interactive techniques, pp. 409–416 (1999)

Lee, C.H., Hu, Y., Selker, T.: iSphere: a free-hand 3D modeling interface. International journal of Architectural Computing 4(1), 19–31 (2006)

Oh, J.Y., Stuerzlinger, W., Danahy, J.: SESAME: towards better 3D conceptual design systems. In: Proceedings of ACM Designing Interaction Systems (DIS), pp. 80–89 (2006)

Schkolne, S., Pruett, M., Schroder, P.: Surface drawing: creating organic 3D shapes with the hand and tangible tools. In: Proceedings of ACM Conference on Human Factors in Computing Systems (CHI), pp. 261–268 (2001)

Google SketchUp (2008), `http://sketchup.google.com` (accessed in May 2009)

Underkoffler, J., Ishii, H.: Urp: a luminous-tangible workbench for urban planning and design. In: Proceedings of ACM Conference on Human Factors in Computing Systems (CHI), pp. 386–393 (1999)

Zeleznik, R.C., Herndon, K.P., Hughes, J.F.: SKETCH: an interface for sketching 3D scenes. In: Proceedings of SIGGRAPH, pp. 163–170 (1996)

Zhai, S., Milgram, P., Buxton, W.: The influence of muscle groups on performance of multiple degree-of-freedom input. In: Proceedings of the conference on Human factors in computing systems (CHI), pp. 308–315 (1996)

# Part V
# Case Studies

Spaces of Design Collaboration
Bharat Dave (The University of Melbourne)

Modeling of Buildings for Collaborative Design in a Virtual Environment
Aizhu Ren and Fangqin Tang (Tsinghua University)

An Immersive Virtual Reality Mock-Up for Design Review of Hospital Patient Rooms
Phillip S Dunston, Laura L Arns, James D Mcglothlin, Gregory C Lasker and Adam G Kushner (Purdue University)

The Immersive Virtual Environment Design Studio
Marc Aurel Schnabel (Chinese University of Hong Kong)

# Spaces of Design Collaboration

Bharat Dave

The University of Melbourne, Australia

**Abstract.** The spread of networked computing environments has led to the development of digital tools and environments to support collaborative design activities. Based on review of our past virtual design studios that employed such tools, this chapter emphasizes the socially and spatially situated nature of collaborative design activities and settings, and identifies critical issues for future development of collaborative virtual environments.

**Keywords:** design studio, virtual teams, situated interactions.

## 1 Virtual Design Studios

The development of the World Wide Web (WWW) in 1991 marked a turning point in reorganization of design practices. Prior to that, the specificities of geographic reach and temporal cycles largely defined how design practices delivered their services, clients they served, consultants they collaborated with, and sites they transformed. The Web collapsed distances, shifted time, rearranged professional dependencies and fundamentally reconfigured spaces of practice for design professionals.

The early sign of changes that eventuated from the convergence of networked information and communication technologies became visible in many places. The research community focused its attention on 'computer-supported cooperative work' (CSCW) with the first international conference on the topic held in 1984, the technology sector echoed with 'network is the computer', and the design institutions embraced virtual design studio experiments that linked dispersed teams of design students via the Web working together on shared design briefs. A review of literature on distributed or virtual design studio experiments over the past two decades reveals rapid development and availability of complementary tools of communication, which include whiteboard, audio and video conferencing, document sharing, textual chat, file transfer and distributed databases, and others. Although many of these tools were developed separately in response to different contexts, the remarkable protocols of the Web for unique and easy addressing, access and display of information among distributed host machines shaped the emerging landscape of collaborative virtual environments.

Supported by a range of synchronous and asynchronous channels of communication and information sharing tools, the early research in virtual studios emphasized

X. Wang & J.J.-H. Tsai (Eds.): Collaborative Design in Virtual Environments, ISCA 48, pp. 143–151.
springerlink.com 

group work as a key aspect of design (Figure 1). In contrast to the preceding developments in design computation that focused on the immediate activities of an individual designer, the flow and sharing of information between networked design studios enabled by the Web led to renewed interest in understanding and supporting collaborative design (Wotjowicz, 1995; Tan and Teh, 1995; Peng, 2000).

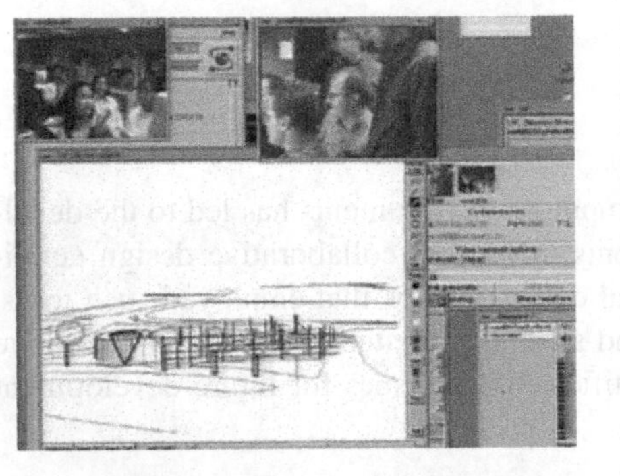
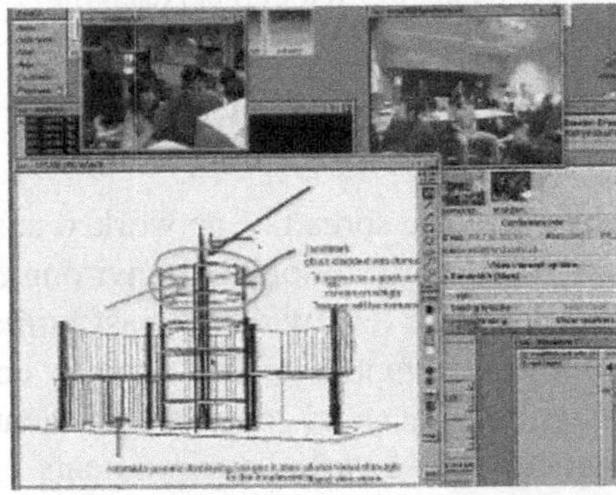
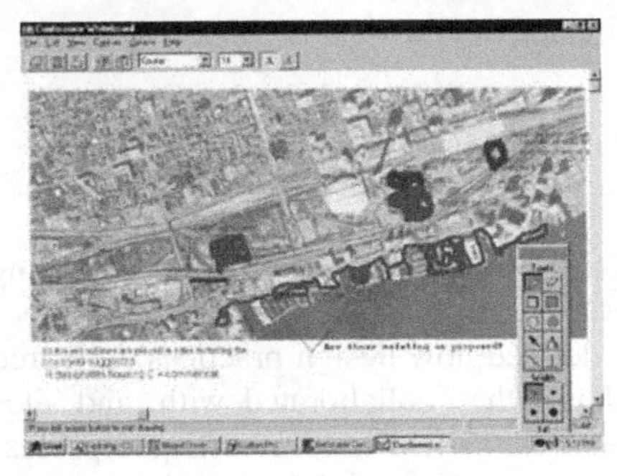
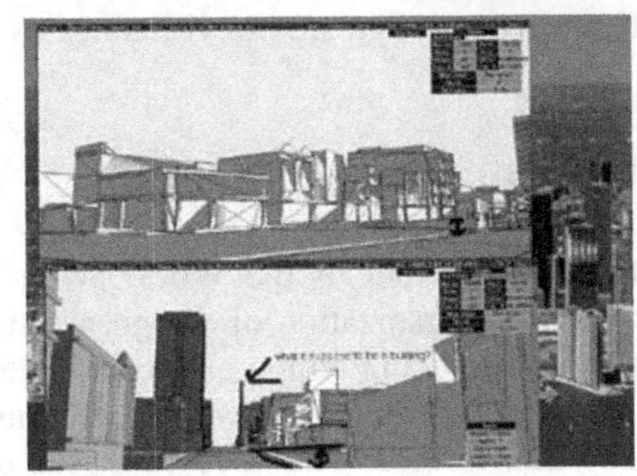

**Fig. 1.** Virtual design studio sessions

Although the Web made visible and gave a particular form to emergent collaborative spaces, such environments were already conceptualized earlier, a few of them anticipated some decades ago even before the advent of the Web. For example, Evans (1969) described one of the earliest group aware environments in the form of an "automated conference room" comprising separate consoles with input devices oriented to four different participants. The networked computer devices supported individual access to information including note taking, sketching, and information retrieval which were mirrored on all the other consoles. Outside of the research laboratories, professional design practices began to employ ad-hoc networks of computers using packet switching technologies and dedicated telephone lines for communication. Although these networks may appear rudimentary by today's standards, the early networking technologies led to functionally networked and distributed design offices. By 1985, for example, Hellmuth Obata Kassabaum, USA, implemented sharing of design documents between its five regional offices using dial-up and dedicated leased lines (Kemper, 1985). The subsequent convergence of information and communication technologies accelerated interest in the need to understand better the nature of collaborative transactions and the settings in which they occur. It enlarged the discussion about virtual environments from being limited to just technical means to also include subtle dimensions of domain-specific exchanges and settings in which collaborative acts unfold.

## 2 Spaces of Divides: Digital and Physical

A number of virtual design studios conducted over the last two decades led us to reflect elsewhere (Dave and Danahy, 1998) on how technologies used, structure of design teams, and patterns of collaboration impact each other. Two key aspects of the nature of electronic collaboration from these early experiments are worth re-emphasizing here.

The early digital design applications evolved around the needs of a single user who worked on a single task at one time (Figure 2, left). Such applications developed efficient and expressive geometric representations and operations matched with a range of interaction functionalities. These applications were constrained as much by available computing technologies (i.e. largely single processor machines with limited memory and speed) as by understanding of designing as an individual, problem-solving process. The focus on single user- single task -single processor changed subsequently in synchronous collaborative environments in which others are invited into a collaborative dialog. The reformulation of a single user design space into a group space for real and virtual collaboration (Figure 2, right) immediately highlighted issues such as priority (e.g. floor control) and recording (e.g., versioning and persistent save of edits) of ongoing design exchanges. Early assumptions about simply reusing single user applications for group work using peer-to-peer, multicasting or other protocols were at best simplistic, at worst clumsy and often frustrating.

**Fig. 2.** Single user and group workspaces

Once the early digital collaborative environments were used for a length of time, some of their other constraints also became evident. Such environments simultaneously provided multiple channels of communication, e.g., talk, text or graphic, and facilitated dense pockets of workspaces typically around a small display screen. In these early experiments, the significance of the broader social and spatial context of real design studios in which design collaborations occur or the fact that collaborative exchanges formed part of a broader workflow was not fully appreciated. The early electronic workspaces existed largely in total isolation from the surrounding work environments. In our collaborative design sessions, we soon realised that the emphasis on display window as the sole target of attention and communication adversely impacted on shared understanding. Since these electronic spaces (even if collocated) existed in physical isolation from each other, they fostered information exchange and understanding that were inscribed by the shape and location of the display surface to a large extent. Although these electronic spaces facilitated spaces of collaboration, they unwittingly also turned into spaces of divides. These observations lead us to suggest that digitally mediated collaborative design spaces need to incorporate some important cues and attributes found in physical design studios that make them spatially and communicatively richer spaces. To this end, it is worth reflecting upon the nature of physical design studio spaces.

Traditional architectural design studios are populated with tangible objects, visible and invisible markers of occupation, authority and work; some spaces that are dedicated and reserved for single purpose and others that accommodate multiple uses; spaces in which specific professional norms and cultures guide behaviors and practices. Such rich complexity of practices and settings is enhanced with a range of information carriers: documents, notes and memos, drawings, models, and other objects. Some of these artifacts such as drawings are prominently visible and pinned up on walls, whereas some others may be rolled up and stored according to perceived patterns of reuse and access but always ready-to-hand. Furniture, display surfaces, patterns of occupancy, orientation, visibility, proximity and social cues, and many other factors lend a dynamic and peculiar character to how the design work- individually and collaboratively gets transacted in studios. Different task contexts lead to different combinations of tangible objects and spatial settings, what may be called *spaces of practice* in which task, artifacts, actors and spaces interact in complex dialectical relationships with each other (Dave, 2003). Many design studio environments are characterized by few formal spaces but abundant presence of overlapping and interacting spaces with amorphous or porous boundaries. For example, the following describes the feel of a celebrated architect's office (Garofalo and Eisenman, 1999): "Entering the office gives you a strange feeling. There are no filters or corridors; you immediately find yourself in a chaotic space, without partitions … In the large open-space office, everyone works together. It does not feel like a real architect's office, but almost like a university lecture room" (p.23). Further, "One need only look around to understand the importance of the three-dimensional control of space: models invade the whole office and the only drawings are the digital elaborations of the complex diagrams used to guide the modeling process" (p.26-27).

The most significant feature of studio environments described above is that collaborative work practices in design studios extend beyond the edges of the drawing board and draw in many objects, places and other people. To develop future collaborative virtual environments that extend beyond the display surfaces into spaces of design studios with porous seams in between needs further investigation of many issues, three of which are discussed next.

## 2.1 Scaffolds and Settings

A steadily growing literature on designing and designers characterizes design studios as variegated, amorphous, and yet purposeful spaces (Bucciarelli, 1998; Cross et. al. 1996). The studio space is a complex web of people and their roles, spaces with multiple functions, artifacts that support designing, exchange and staging, and includes a range of other support infrastructure. Although drawing surfaces occupy central and large portion in design studios, there is a medley of sketches, scale models, material samples, photographs, posters, notes, manifestos, and a myriad of other documents arranged strategically or in close proximity. As Henderson (1995) notes the role of such objects is to express, develop, detail, communicate and present evolving design ideas. As Latour (1986) argues these objects simultaneously support constructing an artifact and staging its performance and understanding by others in a way that it invites others into a dialogue. They act as scaffolds for developing shared understanding and dialogues specific to professional cultures, negotiated and evolved over variable spans of time. Hence some of those scaffolds are persistent whereas others turn out to be ephemeral.

Immersed in an assemblage of visible and easily retrievable, ready-to-hand information artifacts in design studio spaces, designers oscillate between solitary and group work, often reaching over to colleagues for informal and formal exchanges and consultations. It is through these individual and group work practices that a design project evolves through shared project documents that are displayed, retrieved, exchanged and made use of in design studios (Buscher et al., 2001) using multiple, often redundant information representations, something that has so far eluded realization in electronic workspaces.

The design studios facilitate information transactions using various media and representations. These transactions depend upon specific social and spatial settings including roles and responsibilities, arrangement of spatial layout and furniture, display surfaces, patterns of occupancy, proximity and social cues among co-workers and others (Figure 3). No one uniform or universal pattern of work practice dominates within design studios but they are contingent and constantly evolving, adapting and changing as a function of the particular dynamic of people, places, and projects. The nature of objects and interactions in design studios is amorphous and has no definite boundaries. For example, where exactly does design work get done? Everywhere and nowhere in particular though one might point to a drafting table or a group discussion space to locate these activities in a studio.

Taking cues from traditional studio settings, future workspaces for individuals and group interaction need to flow beyond fixed displays into spatial surroundings, an agenda that appears to be implicit in projects such as 'The Media Space' at Xerox Parc (Harrison, 1993), and more recently in projects such as Interactive Workspaces (Fox et. al. 2000), Tangible Media (Ishii et. al. 2002), Roomware (Streitz et. al. 2001), BlueSpace (Chou et. el., 2001), immersive spaces (Dave, 2001).

**Fig. 3.** Studio artifacts and settings

## 2.2 Synoptic View

Digital modelling tools involve representation of design projects around a single digital database from which specific views are generated, displayed, edited and exchanged. Although only a partial view of information may be displayed at any given moment, digital representations of design models contain far more information internally. A single view of a project on a finite display surface does not fully reveal all the other information that may be embedded in the same model. The imposed seriality of information display, i.e., one view replaced by another on the same display surface, may foster fragmented understanding of project information. The traditional design studios, in contrast, employ a range of media and representations displayed simultaneously including drawings, photographs, models, etc. The use of sometimes redundant and parallel representations in traditional studio discussions enables a holistic reading and cross-referencing of design projects. This is not easy to achieve with the use of digital workspaces in which only one or a few views of the project are visible at any given time. The partial and fragmented views of digital information need to be complemented in future collaborative workspaces with simultaneous access to holistic and synoptic views.

## 2.3 Foci of Interaction

The display surface and interaction tools developed for single users may be limiting for supporting group interactions. When designers gather traditionally around a table with models or drawings, discuss and occasionally draw or move things around, a number of actions unfold in parallel, all *centred* around a visible information object such as drawings using a spatial reference to further indicate the locus for dialog. This may be in the form of gestures, pointing to a specific section of drawings, or taking apart a model. Such actions are quite transparent to others, there is no explicit need to declare a tool to be used, action to be performed, or location where it is to be carried out. It happens as one fluid interaction in which eyes, hands, location, and object manipulations complement each other. In contrast, similar interactions in electronic workspaces appear to be become *decentred* since they involve bounded and isolated digital displays, input devices, information sources, and users located in disjoint spaces. It may happen due to the location and orientation of display surfaces, or the need to know who initiates the next action while also maintaining some eye contact, watching gestures of others, and keeping an eye on the changing information, all of which may be spatially dispersed or discontinuous. As Tory et al. (2008) emphasize, these issues assume even greater significance in synchronous collaborative work settings.

# 3 Reaching Out

The issues discussed above- memory scaffolds and settings, need for synoptic views of project information and means to embed spatial interactions, are just three of the major challenges for future virtual workspaces for individual and group work practices. Existing practices provide some clues about what may be useful in future workspaces but there is no a priori theory we can draw upon to this end. Availability and adoption of any kind of digital tools alters the very design practices they are intended to serve.

In the face of a moving research target, it is apt to recall Grudin's (1990) characterization of the evolving foci of interactive systems development: at the hardware (1950s), at the software (1960-70s), at the terminal (1970-80s), at the interaction dialogue (1980s), and at the work setting (1990s). The reach of computing has been expanding from the workspaces of individual users to encompass group work settings. As the reach of collaborative environments increases in the coming decades and becomes embedded in routine work practices, new typologies of work have emerged (Kimble et al., 2000) firstly along dimensions of time and place, and increasingly along the third dimension of organization (Figure 4). Each cell in this typology revolves around a different combination of tools, representations and actors situated in different social and spatial settings and roles, and located somewhere along the larger spectrum of project work flow and culture of practice.

The issues of collaborative virtual environments are not simply technical problems. The situated nature of collaborative design practices requires that digital environments to support them are considered as an important but just one of the many constitutive components of the social and spatial context of design studios, and not something that can be studied or ought to be supported in isolation.

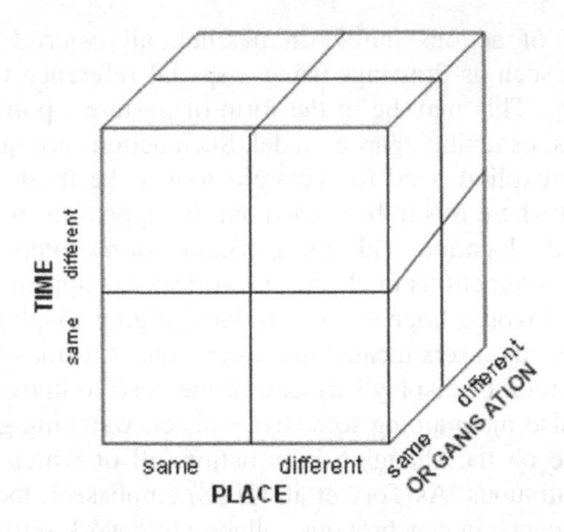


**Fig. 4.** Dimensions of collaborative practices

## Acknowledgements

The design studios referred to in this chapter were conducted over the last few years and are described in detail elsewhere (Dave, 2003; Dave, 2001; Dave and Danahy, 1998; Dave, 1995).

## References

Bucciarelli, L.L.: An ethnographic perspective on engineering design. Design Studies 9(3), 155–168 (1988)

Buscher, M., Gill, S., Mogensen, P., Shapiro, D.: Landscapes of practice. The Journal of Collaborative Computing 20(1), 1–28 (2001)

Chou, P., Gruteser, M., Lai, J., Levas, A., McFaddin, S., Pinhanez, C., Viveros, M., Wong, D., Yoshihama, S.: BlueSpace: Creating a Personalized and Context-Aware Workspace (2001)

Cross, N., Christiaans, H., Dorst, K. (eds.): Analysing Design Activity. Wiley, Chichester (1996)

Dave, B.: Hybrid spaces of practice. In: Chiu, M., Tsou, J., Kvan, T., Morozumi, M., Jeng, T. (eds.) Digital Design: Research and Practice. Proceedings of the 10th International Conference on CAAD Futures 2003, pp. 181–190. Kluwer Academic Publishers, Doordrecht (2003)

Dave, B.: Immersive modelling environments. In: Jabi, W. (ed.) Reinventing the Discourse: Proceedings of the 21st Annual Conference of the Association for Computer-Aided Design In Architecture, Buffalo, New York, pp. 242–247 (2001)

Dave, B., Danahy, J.: Virtual study abroad and exchange studio. Automation in Construction 9(1), 57–71 (2000)

Dave, B.: Towards distributed computer-aided design environments. In: Tan, M., Teh, R. (eds.) The Global Design Studio: Proceedings of the Sixth International Conference on CAAD Futures 1995, Centre for Advanced Studies in Architecture, pp. 659–666. National University of Singapore, Singapore (1995)

Evans, D.: Augmenting the human intellect. In: Milne, M. (ed.) Computer Graphics in Architecture and Design, Yale School of Architecture and Design, New Haven, pp. 61–66 (1969)

Fox, A., Johanson, B., Hanrahan, P., Winograd, W.: Integrating Information Appliances into an Interactive Workspace. IEEE CG&A, Los Alamitos (May/June 2000)

Garofalo, L., Eisenman, P.: Digital Eisenman: An Office of An Electronic Era. Birkhauser-Publishers for Architecture, Boston (1999)

Grudin, J.: The computer reaches out: The Historical continuity of interface design. In: Chew, J.C., Whiteside, J. (eds.) CHI 1990 Conference Proceedings, Empowering People, pp. 261–268 (1990)

Harrison, S.: CAD in research: Computing and the social nature of design. ACADIA Quarterly Newsletter 12(1), 10–18 (1993)

Henderson, K.: The Visual Culture of Engineers. In: Star, S.L. (ed.) The Cultures of Computing, pp. 197–218. Blackwell, Malden (1995)

Kemper, A.M. (ed.).: Hellmuth Obata Kassabaum. In: Pioneers of CAD, pp. 172–186. Hurland/Swenson Publishers, Pacifica (1985)

Kimble, C., Li, F., Barlow, A.: Effective Virtual Teams through Communities of Practice. Strathclyde Business School, Research Report 2000/9 (2000)

Latour, B.: Visualisation and cognition: thinking with eyes and hands. In: Studies in the Sociology of Cultures Past and Present, pp. 1–40 (1986)

Peng, C.: Design Through Digital Interaction. Intellect, Oxford (2000)

Rekimoto, J., Saitoh, M.: Augmented surfaces: A spatial continuous workspace for hybrid computing environments. In: Proceedings of CHI 1999 Conference on Human Factors in Computing Systems, pp. 378–385. ACM Press, CA (1999)

Streitz, N., Tandler, P., Müller-Tomfelde, C., Konomi, S.: Roomware: Towards the next generation of human-computer interaction based on an integrated design of real and virtual worlds. In: Carroll, J. (ed.) Human-Computer Interaction in the New Millennium, pp. 553–578. Addison-Wesley, Reading (2001)

Tan, M., Teh, R. (eds.): The Global Design Studio: Proceedings of the Sixth International Conference on CAAD Futures 1995. Centre for Advanced Studies in Architecture. National University of Singapore, Singapore (1995)

Tory, M., Staub-French, S., Po, B.A., Wu, F.: Physical and digital artifact-mediated coordination in building design. In: Computer Supported Cooperative Work (CSCW), vol. 17(4), pp. 311–351. Springer, Heidelberg (2008)

Wojtowicz, J. (ed.): Virtual Design Studio. Hong Kong University Press (1995)

# Modeling of Buildings for Collaborative Design in a Virtual Environment

Aizhu Ren and Fangqin Tang

Tsinghua University, Beijing, China

**Abstract.** The application of virtual reality systems to the civil and building engineering shows that a great deal of work load is the modeling of a building, or buildings in an urban area. To provide a modeling system which enables the users to construct models of irregular and complicated structures efficiently, and to share the models via network for the design and construction management, a modeling system of application independent, which enables quick modeling of irregular and complicated building structures adapted to VR applications based on Web was developed. The 3D building model can be transferred to the model which can be viewed in different virtual reality environments via a special interface for the data conversion. In the constructing of 3D models of urban objects for urban applications, the digital maps of urban area were used to reduce the work loads.

**Keywords:** building, CAD, collaborative design, modeling, virtual reality.

## 1 Introduction

VR (Virtual Reality) is a technology that generates simulative environment of real world by computers. In this environment, through different kinds of sensors, users are involved in such an artificial virtual environment and can interact with it directly and naturally. Because of the lively expressive ability of the VR technique, the communications among designers, constructors and owners become easier. This technique has therefore been employed in the planning, design and construction management of buildings, and the simulation of disaster damages.

A Chinese museum was designed based on virtual reality (Liu, 2008) in which Virtual Reality Modeling Language (VRML) was employed. VRML is an international 3d modeling standard, designed particularly for web applications. VR and GIS are employed in the urban planning of Huangdao district (Han et al., 2007) of 275 $km^2$ area, in which 9000 models (buildings, roads, trees, flowers, etc.) over 20 $km^2$ were built. A construction management system was developed for the construction simulation of the main stadium of 2008 Olympic Games in Beijing and Qingdao bridge based on OpenGL (Hu et al., 2008). CHEN Chi studied on the application of virtual reality-based system for the fire fighting and emergency response in an underground station (Ren et al., 2006). A GIS, CAD,

X. Wang & J.J.-H. Tsai (Eds.): Collaborative Design in Virtual Environments, ISCA 48, pp. 153–165.
springerlink.com 

FEA and VR integrated system for the simulation of building damages due to earthquake was developed (Xu et al., 2008).

The experience in the application of virtual reality systems to the civil and building engineering shows that a great deal of work load is the modeling of a building, a bridge or buildings and roads in an urban area. For instance, 3 persons worked for 3 weeks to construct a computer model for the steel structure, and 3 persons worked for 1 week to construct a computer model for the reinforced concrete structure of the main stadium of Beijing 2008 Olympic Games (Hu et al., 2006), since there are 9200 components which consist of more than 3,500,000 triangles for the construction simulation. More than 30 persons in one year are required to construct the models into the computer for more than 80,000 apartments and buildings for urban management applications (Ren et al., 2004).

# 2 Modeling of Buildings for Collaborative Architectural and Structural Design

Computer models of buildings for the architectural design can be constructed with many existed software such as AutoCAD, 3ds Max, Revit, SketchUp etc. The models constructed by some software such as 3ds Max can be directly converted to VRML models, so that the model can be viewed in a virtual reality environment.

Computer models of buildings for the structural design can also be constructed with many existed structural analysis software such as SAP, ANSYS, MSC.MARC, etc. The models constructed by structural analysis generally can not be converted directly to VRML or other models which can be viewed in a virtual reality environment.

The building structures have become more and more complicated, since many advanced technologies have been used in the building industry. To efficiently construct the models of irregular and complicated structures, and to share the models among different offices and companies via network for the design and construction management, a system with the following features was developed (Ren et al., 2004):

(1) Enables the quick modeling of irregular and complicated building structures such as lofting operations;

(2) Realizes the conversion from the CAD model to a XML-based neutral model independent of applications or platforms;

(3) Implements the conversion interface which can adapt the neutral model into VR applications, e.g. VRML applications.

## 2.1 Modeling of Abnormity Buildings in CAD Environment

An irregular building structure usually has a large amount of components, which may have different geometry or locations. Quick modeling for these complex structures requires a good understanding of the component features both on geometric shapes and spatial distributions. Thus the operation mode for modelers can be designed accordingly. A modeling system was developed based on ObjectARX library to facilitate the modeling of abnormity buildings in CAD environment.

On the abstraction of the component features, class structures are designed to respectively represent the physical components, which include wall, beam, column and slab, as shown in Figure 1. Derived from the AcDbEntity class, each class can customize its own properties and methods and save the relevant data into AutoCAD database. In addition, with the extensibility of object-oriented design, new entity classes can be defined for specific components in certain building structures.

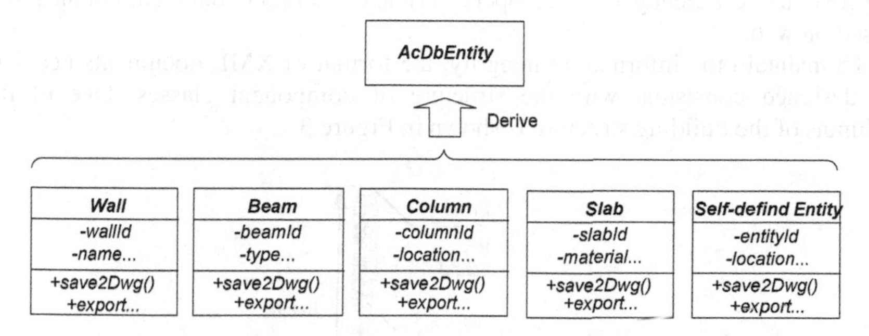


**Fig. 1.** The hierarchy of component classes

To improve the efficiency of 3d modeling, a lofting function was developed in which an arbitrary organized component group can be cloned and placed along a specific path with variable scales, thereby simplifying the modeling process. Two operation modes, i.e. "uniform lofting" and "variable scale lofting" are provided to modelers. In the uniform lofting, the components are entirely cloned and added at certain locations of the scene without any variations in lengths or shapes. In the variable scale lofting, the components are placed along different track curves separately. Some of the component lengths, therefore, may vary according to the lofting path.

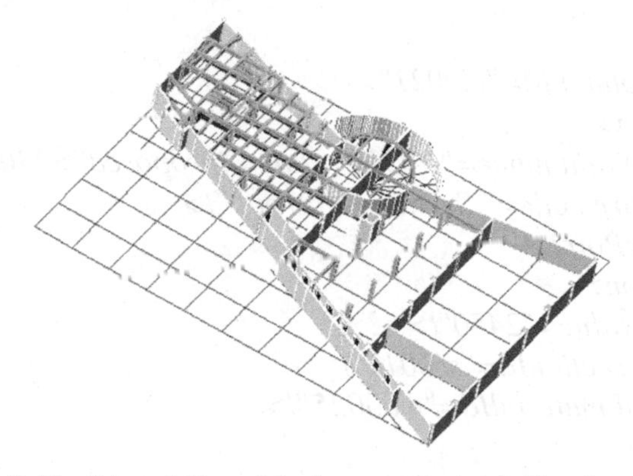

**Fig. 2.** The 3d modeling of the basement using lofting functions

A computer model was constructed with the modeling system for the design of Beijing Seine-Villa public house, which is a four storey building with irregular shape. Figure 2 indicates the 3d modeling process of the basement using lofting functions.

## 2.2 The XML-Based Neutral Model

The XML (eXtensible Markup Language) standard was selected to construct a neutral model for the information sharing between applications. As an extensible specification to organize, store and transport information, XML can be used to improve the efficiency of data operations and facilitate data communications based on web.

To maintain the information integrity, the format of XML documents needs to be designed consistent with the structure of component classes. One of the columns of the building structure is shown in Figure 3.

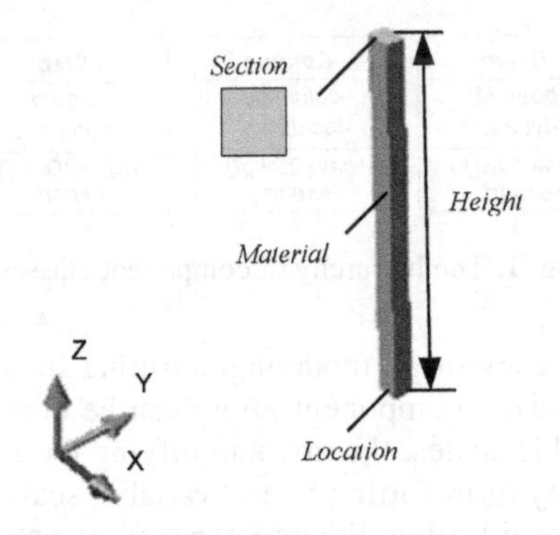


**Fig. 3.** A column of the building structure

The column class defines the geometric attributes, e.g. *Height*, *Section* and *Material*, and the spatial attributes, e.g. *Location*, to represent the features of columns. A XML-based neutral model document was designed accordingly as follows.

```
<Column columnId="cl0021">
  <Location>
    <BasePoint name="PT310" state="proposed">2300 4700 3100
    <Feature code=" Section Centroid "/>
    </BasePoint>
  </Location>
  <Height value="2450"/>
  <Section sectionId="sc0012"/>
  <Material materialId="ma0025"/>
</Column>
```

The set of tags and the schema created in the XML document will be utilized to standardize the treatment of model information, such as the presentation of model data, data transfer, or the communication between applications (White et al., 2002). Since a building structure consists of a large amount of components, the document usually contains a huge number of nested elements and its size can be quite large. A typical XML document structure for an entire building structure is as follows.

```
<BuildingStructure>
  <BuildingStorey buildingStoreyId="bs0001">
    <Beam beamId="bm0032">
      ......
    </Beam>

  </BuildingStorey>
  ......
</BuildingStructure>
```

The elements such as beams and columns declared in the XML document describe the components by stories and thereby formulate the entire building structure. An export function was customized for each component class and outputs the class attributes according to the predefined XML formats.

The component entities in different stories are stored by layers in CAD model. An ARX-based conversion module then visits each CAD layer, traverses the entities on the layer and calls the export function of the related class to append the component elements to the neutral model. The elements for the components within one same storey are placed inside one *BuildingStorey* element. The conversion process from CAD model to XML-based neutral model is therefore established.

Since the lofting functions are provided, the modeling of irregular buildings becomes easier. The conversion module makes it possible to automatically generate the neutral model corresponding to the lofting results, and with the appropriate interfaces, the neutral model can be freely converted into different environments, therefore increasing the reusability of the modeling work. Meanwhile, the strength of XML in data management and transportation makes it efficient for the information exchange related to the building models through both local access and web-based communications.

## 2.3 Generation and Visualization of Corresponding Virtual Reality Models

The neutral model makes it possible to reconstruct the 3d model in independent applications and environments. Normally two steps are needed for the reconstruction. First, the semantic analysis for the neutral model must be performed to extract the building information. Second, the interface needs to be designed and implemented to output the corresponding model compatible to new environments. A conversion process was designed according to these two steps to obtain the VRML model from the neutral model:

(1) Interpret the data of neutral model by a XML document parser developed on SAX (Simple API for XML);

(2) Export the analysis results into VRML formats through the visit of a conversion interface.

Derived from the base class of XML document handler, a class named XMLNeutralDocParser specifies the operations performed during the handling. One of the advantages of using SAX is that the developers can customize the parsing process only by overriding the virtual functions. The virtual function startElement() was overridden to read the attributes of the elements, while the function named characters() is responsible for the extraction of the text data inside the elements. These virtual functions constitute the interfaces which will be driven by specific events. The semantic analysis then can be performed with a series of events triggered during XML document reading.

Based on the results of semantic analysis for the neutral model, the conversion interface can export the building information in VRML formats. Some calculations and transformations are needed, however, since the geometry may be described differently in two sets of formats. Take the column in Fig. 3 as an example. The following codes indicate the corresponding VRML model generated from the neutral model:

```
DEF cl0021 Transform {
    translation 2300 4700 3100
    children [
        Shape {
            appearance Appearance {
                material Material {
                    diffuseColor 0.498 0.749 1
                    ambientIntensity 1.0
                    specularColor 0 0 0
                    shininess 0.525
                    transparency 0
                }
            }
            geometry Box {
                size 600 600 2450
            }
        }
    ]
}
```

Once the VRML model is created, it can be roamed interactively in Internet Explorer with appropriate plug-ins such as Cortona VRML Client (Zhao et al.,

2006). With the advantage of VRML in web applications, it can also be viewed either locally or distributed across the Internet (Whyte et al., 2000). Figure 4 shows that the building model of Beijing Seine-Villa Public House which is constructed by the modeling system has been converted into a VRML model through the XML-based neutral model.

**Fig. 4.** Navigation of the VRML building model

In summary, a modeling system was designed to facilitate 3d modeling in CAD environment and generate the corresponding VRML model through a XML-based neutral model, as shown in Figure 5. Besides the VRML model, furthermore, it is also possible to acquire other forms of models with appropriate conversion interfaces oriented to the related formats. For example, for Vega applications, the OpenFlight API can be employed to convert the neutral model into FLT format, which can be automatically loaded into Vega scenes.

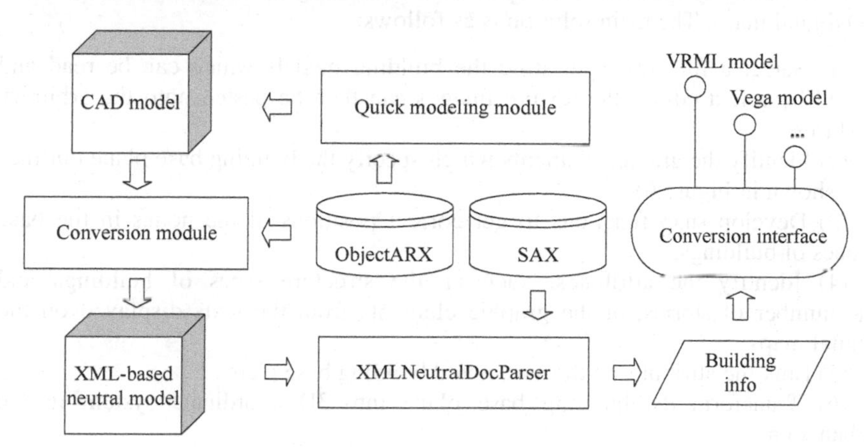


**Fig. 5.** The system operation flow

## 3 Modeling of Buildings in an Urban Area

Three-dimensional models of urban objects play an important role in the urban applications such as urban planning, environmental concerning, or urban disaster mitigations. However, the modeling of urban objects is time consuming. It is therefore necessary to obtain simple but adaptable building model for applications in urban area. In order to solve the significant problem for the modeling of a mass of buildings for the development of digital disaster reduction system, a new method for the construction of a mass of buildings and road networks from the current available digital maps for urban area is necessary.

### 3.1 Automatic Model Generation from a Digital Map of an Urban Area

In general, it is not necessary that every building model should be constructed in details in many urban applications, such as landscape analysis, wind simulation, or flood simulation for urban area. The modeling costs for urban area in China are significantly expensive due to the big population and in turn the dense covering of buildings and roads. For instance, there are more than 1,000,000 peoples in the most middle size cities in China. In turn there are about 80,000 apartments and buildings, and other urban objects in those urban areas. More than 30 persons in one year are required to construct the models into the computer for more than 80,000 apartments and buildings by employment of current available software. It is necessary to work out simple but adaptable building model for urban applications. Since two dimensional (2D) digital maps were drawn by AutoCAD which are utilized popularly in most urban areas in china, it is reasonable to construct 3D models of buildings directly from those 2D digital maps, in which the base plane outlines, the plane geometry dimensions and the number of floors of buildings can be identified from the 2D maps. This saves significantly the modeling costs for buildings with regular planes and elevations.

A modeling system was developed for constructing 3D models based on urban 2D digital maps. The main solution is as follows:

(1) Select a file format to store the building models which can be read and edited by a text editor, the texture images can then be pasted onto the arbitrary surfaces.

(2) Identify the graphic elements which specify the building base plane outlines (As shown in Figure 6).

(3) Develop snap functions to get correct positions of the nodes in the base planes of buildings.

(4) Identify the attributes, such as the structure types of buildings and the number of stories, of the graphic elements from the texts displayed on the digital maps.

(5) Link the attributes to the associated building base plane.

(6) Transform the building base plane into 3D coordinate system in the urban area.

(7) Construct the theoretical 3D models for the buildings. The heights of the buildings are calculated from the attribute data: the number of stories and the building type. For instance, the story height is taken as 2.9m for the apartments.

(8) Construct the practical 3D models for the buildings in urban area in which each surface in the theoretical models are shrunk in a small value from the intersection edges, so that each surface in a practical building model is independent of the building model.

(9) The surfaces in a building model are numbered in a sequence. While the texture images associated to the surfaces are numbered in the same sequence. The texture images can then be pasted automatically onto the associated surfaces (As shown in Figure 7).

**Fig. 6.** A digital map of an urban area

**Fig. 7.** Automatic texturing to the building

For the buildings with irregular shapes and non-plane roofs, a specific modeling system was developed for the constructing of irregular building models. The components in a building are specified in a 3D CAD model which can then be inserted into the urban model. As soon as a 3D CAD model has been constructed, the model is inserted into the urban model according to the base plane outline of the building which is obtained automatically from the digital urban maps.

The building models generated from a digital map of an urban area can be employed in the urban planning and scene roaming. For urban applications such as damage simulation due to earthquake, the specific building models should be constructed.

## 3.2 Building Models for Damage Simulation Due to Earthquake in an Urban Area

The building responses due to a predefined earthquake wave were simulated for an urban area. Based on the building model automatically generated from the digital map, the building details and distributions can be extracted for further calculations. The finite element package MSC.MARC was employed to evaluate the damage level for the more than 7000 buildings existed in this urban area. Figure 8 indicates the evaluation procedure as well as the related parameters. A simplified building structural model with mass concentrated at each floor is generated with necessary parameters such as elastic stiffness. The damage level is then established by the analysis of the structural model.

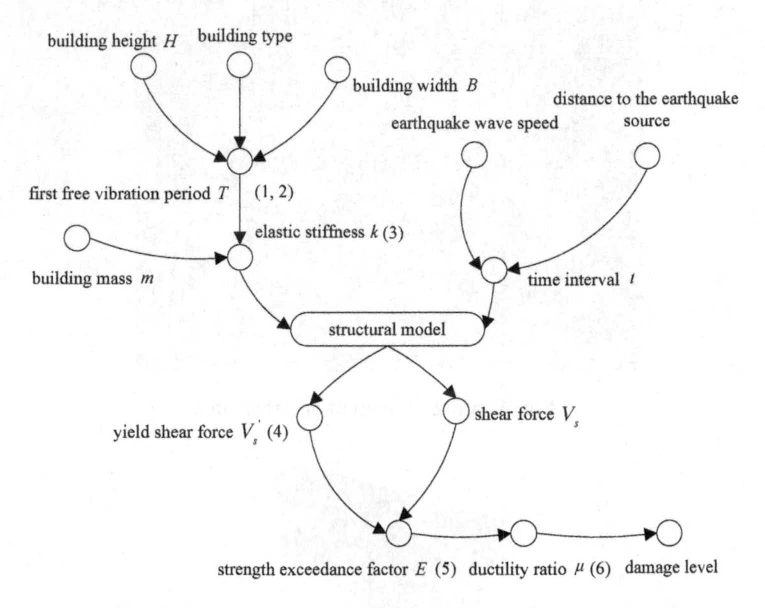


**Fig. 8.** The evaluation of the damage level due to earthquake

The equations numbered in Figure 8 are as follows.

The first free vibration period $T$ is calculated according to the building size and type. For frame and frame-shear wall structures:

$$T = 0.06 + 0.046\frac{H}{\sqrt[3]{B}} \tag{1}$$

For shear wall structures:

$$T = 0.016H \tag{2}$$

The elastic stiffness $k$ will be

$$k = \frac{4\pi^2 m}{T} \tag{3}$$

The yield shear force $V_s'$ for each floor is given by the following equation (Yin, 2004):

$$V_s' = 1.39\alpha W \frac{\sum_s^n W_i i}{\sum_1^n W_i i} \tag{4}$$

The strength exceedance factor $E$ and the ductility ratio $\mu$ are further obtained by

$$E = \frac{V_s}{V_s'} \tag{5}$$

$$\mu = \sqrt{E} e^{1.9(1-\frac{1}{E})} \tag{6}$$

The damage level of a building can be determined by the empirical relationship shown in Table 1 (Yin, 1996).

**Table 1.** Damage level of a building

| Damage level | Not damaged | Slight damaged | Secondary damaged | Serious damaged | Destroyed |
|---|---|---|---|---|---|
| *Frame and frame-shear wall structures* | $\mu \le 1$ | $1 < \mu \le 3.7$ | $3.7 < \mu \le 6$ | $6 < \mu \le 8.2$ | $\mu > 8.2$ |
| *Shear wall structures* | $\mu \le 1$ | $1 < \mu \le 2.0$ | $2 < \mu \le 4$ | $4 < \mu \le 6$ | $\mu > 6$ |

### 3.3 Integration of GIS, CAD and VR View Port

A system which integrates three interactive views, namely GIS view, CAD view and VR view was implemented through the MFC-based framework. The GIS view of the integrated environment provides a variety of operations for spatial data management, the CAD view visualizes the damage level analyzed by the structural model with various colours and the VR view enables the immersive roaming in virtual scenes. Furthermore, each view is interdependently connected, i.e. the viewport transformation in any view will lead to the corresponding changes in others. Figure 9 shows the integrated interface of the system.

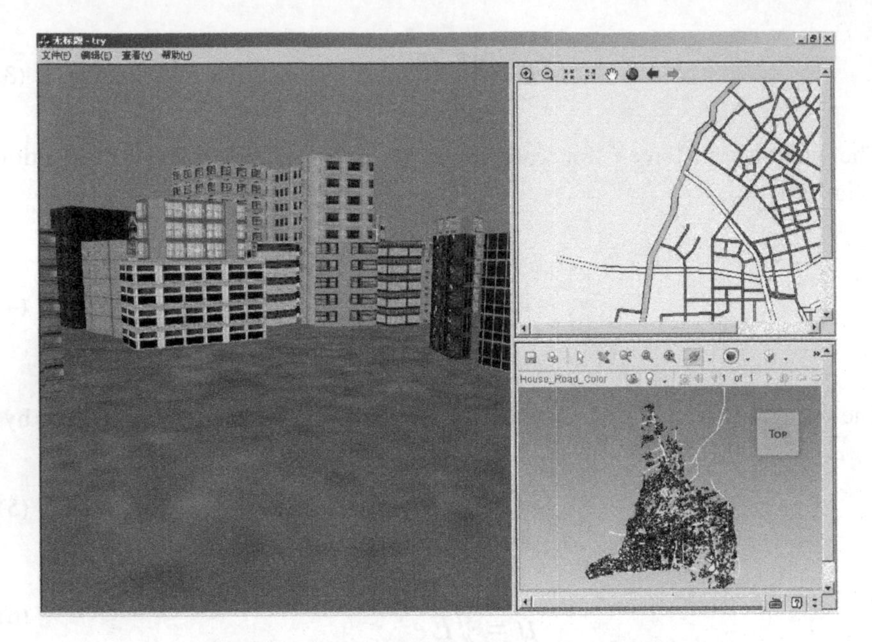


**Fig. 9.** A GIS, CAD, VR integrated viewport

# 4 Conclusions

Construction of computer models for a building or buildings in an urban area for collaborative design in a virtual environment is a big work. The following solutions may reduce the work loads:

(1) Development of a neutral model based on XML may lead to the building model be independent of the applications. Development of interfaces among CAD models and virtual reality environment makes the building models more flexible so that the design or construction process can be displayed in different virtual reality environments as user specifies.

(2) In the constructing of 3D models of urban objects for urban applications, the digital maps of a city can be used if the building base planes are drawn in a correct way. Simplification of buildings with regular shapes and plane roofs is reasonable for the urban applications such as urban planning or disaster simulation.

(3) An integrated GIS, CAD and VR environment may raise the efficiency of the collaborative design in a virtual environment for urban applications.

# Acknowledgements

The work presented in this article were mainly done by my following graduate students under my supervision: Dr. Wen Yang, Dr. Xu Yun, Mr. Chen Xuping

(Master), Dr. Chen Chi, Dr. Pan Guoshuai, Dr. Tang Fangqin, Dr. Shi Jianyong and Mr. Xu Feng. I sincerely thank my students for their hard working.

The research was funded by the National Key Project of Scientific and Technical Supporting Programs from Ministry of Science & Technology of China (2006BAK01A02-09 and 2006BAJ06B06). The authors sincerely thanks for the supports. The authors also thanks the support from Tsinghua University—Hong Kong Polytechnic University, Center for Information Technology in Construction (CITEC).

## References

Han, Y., Qiao, X., Sun, W., Zhang, L.: Application of virtual reality GIS in urban planning - An example in Huangdao district. In: Proceedings of SPIE - The International for Optical Engineering, Nanjing, China (2007)

Hu, Z., Zhang, J., Lu, M., Cao, M., Gao, S.: Simulation and optimization of prefabricated steel-structure installation operations. In: 1st International Construction Specialty Conference, Calgary, Alberta, Canada (2006)

Hu, Z., Zhang, J., Zhou, Y., Zhang, Y.: Research and development of 4d construction management system for Qingdao bay bridge. Construction Technology 12, 84–87 (2008) (in Chinese)

Liu, X., Qiao, J.: Research on Chinese museum design based on virtual reality. In: Proceedings - 2008 International Workshop on Modelling, Simulation and Optimization, WMSO 2008, Hong Kong, China, pp. 372–374 (2008)

Ren, A., Chen, C., Zou, L., Tang, F.: A virtual reality-based system for the fire fighting and emergency response. In: Proceedings - Joint International Conference on Computing and Decision Making in Civil and Building Engineering, Montreal, Canada, pp. 1447–1456 (2006)

Ren, A., Shi, J., Pan, G., Xu, Y., Wen, Y.: Modeling of buildings and roads for urban applications based on 2d digital maps. In: Proceedings - Xth International Conference on Computing in Civil and Building Engineering, Weimar, Germany (2004)

Ren, A., Wen, Y., Chen, C., Shi, J.: Modeling of irregular structures for the construction simulation in virtual reality environments based on web. Automation in Construction 13(5), 639–649 (2004)

White, C., Quin, L., Burman, L. (eds.): Mastering XML. Electronic Industry Press, Beijing (2002)

Whyte, J., Bouchlaghem, N., Thorpe, A., McCaffer, R.: From CAD to virtual reality: modeling approaches, data exchange and interactive 3D building design tools. Automation in Construction 10(1), 43–55 (2000)

Xu, F., Chen, X., Ren, A., Lu, X.: Earthquake disaster simulation for an urban area, with GIS, CAD, FEA and VR integration. Tsinghua Science and Technology 13(S1), 311–316 (2008)

Yin, Z. (ed.): Analysis of Losses Due to Earthquake and Resistance Standard. Earthquake Press, Beijing (2004) (in Chinese).

Yin, Z. (ed.): Earthquake Disaster and Prediction of Losses Due to Earthquake. Earthquake Press, Beijing (1996) (in Chinese)

Zhao, W., Shen, L., Li, M., Zheng, J.: Approach of interactive operation of VRML model in virtual collaborative design. In: Proceedings-10th International Conference on Computer Supported Cooperative Work, Nanjing, China, pp. 1168–1172 (2006)

# An Immersive Virtual Reality Mock-Up for Design Review of Hospital Patient Rooms

Phillip S. Dunston, Laura L. Arns, James D. Mcglothlin, Gregory C. Lasker, and Adam G. Kushner

Purdue University, U.S.A.

**Abstract.** Full scale physical mock-ups of specific hospital units such as patient rooms are routinely utilized to serve the multiple purposes of constructors, designers, and owner stakeholders for healthcare facility projects. The shortcoming with this practice is that the mock-up is constructed during the construction phase and is of limited use for making extensive decisions regarding the functionality of the room design. Three-dimensional visualisation tools offer healthcare facility stakeholders the opportunity to comprehend proposed designs more clearly during the planning and design phases, thus enabling the greatest influence on design decision making. While several options exist, based on their experience with a bariatric patient room model, the authors promote the utilization of Virtual Reality mock-ups for design review because of their enhanced capacity for an immersive, interactive experience with the design and for the long-term utility of such models for the balance of the project life cycle.

**Keywords:** design review, hospitals, immersive display, virtual mock-up, virtual prototype, virtual reality.

## 1 Introduction

Construction of full-scale physical mock-ups (PMUs) during the construction phase of healthcare facility projects is a common practice, serving as a type of submittal which provides both the opportunity for the constructor to practice construction methods and the opportunity for the other healthcare project stakeholders to see the truest representation of the design of key facility units, such as patient rooms, nurse stations, operating rooms, etc. Based on the most recent design, the constructor builds these PMUs to a desired level of completion—anywhere from drywall stage to complete build-out (i.e., all finishes, furniture, equipment, and electrical lighting). The architect and hospital administrators, doctors, nurses, and technicians then inspect the PMU and provide feedback regarding details that may yet be changed before the actual rooms or units are constructed. With the footprint and general layout of the room being set during the earlier design phase, improvements identified from the PMU review are limited to minor space modifications, finishes, some fixtures, and equipment models. Three-dimensional

X. Wang & J.J.-H. Tsai (Eds.): Collaborative Design in Virtual Environments, ISCA 48, pp. 167–176.
springerlink.com 

modeling and visualization enable extension of the benefits of PMU review to the early planning and design stage where greater improvements are possible at less cost.

In particular, Virtual Reality (VR) modeling and display technologies offer the greatest potential to improve the design through early design visualization, better serving the purposes of the designer and the healthcare organization stakeholders. Based on the convincing body of evidence demonstrating that the physical environment in healthcare facilities impacts patient recovery, patient and staff safety, and the quality of patient care, practitioners are promoting design for healthcare facilities that is guided by rigorous research linking the physical environment of hospitals to patient and staff outcomes (Hamilton, 2003; Hamilton and Watkins, 2009). This practice of *evidence-based design* can be facilitated by the immersive and interactive 3D visualization that constitutes VR mock-ups.

What follows in this chapter is a rationalization for the utilization of VR mock-ups and a description of case study experience with a VR mock-up of a patient room. A brief review of other design phase visualization methods is presented to provide perspective on the capabilities of the VR mock-up. Lessons learned in the course of developing and evaluating the VR mock-up and ideas for useful feature extensions are also presented.

## 2 Rationale for Utilizing Mock-Ups in the Early Design Stage

As proposed by Eastman et al. (2008), a paradigm shift in a long-understood project characteristic helps to comprehend the potential for impacting the overall project by incorporating virtual (computer-represented) mock-ups during the design phase. Figure 1, illustrates this potential by overlaying conceptual models of design decision influence (Line 1), cost of implementing design decisions/changes (Line 2), and the relative timing of effort involved in completing the project design (Line 3). Line 1 depicts the well-known fact that design decisions made earlier in the life cycle of a project have the least cost associated with their implementation and yet the greatest influence on determining the constructed facility's functionality and downstream project costs. Design changes enacted later in the project, as indicated by Line 2, are more costly because constraints imposed by earlier design decisions result in greater disruption (i.e., rework and engineering and administrative delays) to the project.

The key motivation for utilizing virtual mock-ups is illustrated in Figure 1 by repositioning Line 3 as indicated by dashed Line 3′. By utilizing virtual modeling—in this case, VR mock-ups—to enable better design decisions sooner in the project, greater impact on the facility's functionality is achieved. This concept is illustrated by Line 3′ being underneath the design decision influence Line 1. Also, the cost of design changes is minimized. While Eastman and his colleagues applied this notion to the utilization of building information modeling (BIM), similar impact can be expected from any process or tool that enables the finalization of design decisions earlier in the project life cycle. The common practice of waiting until the construction phase to finalize decisions via PMUs misses opportunities to

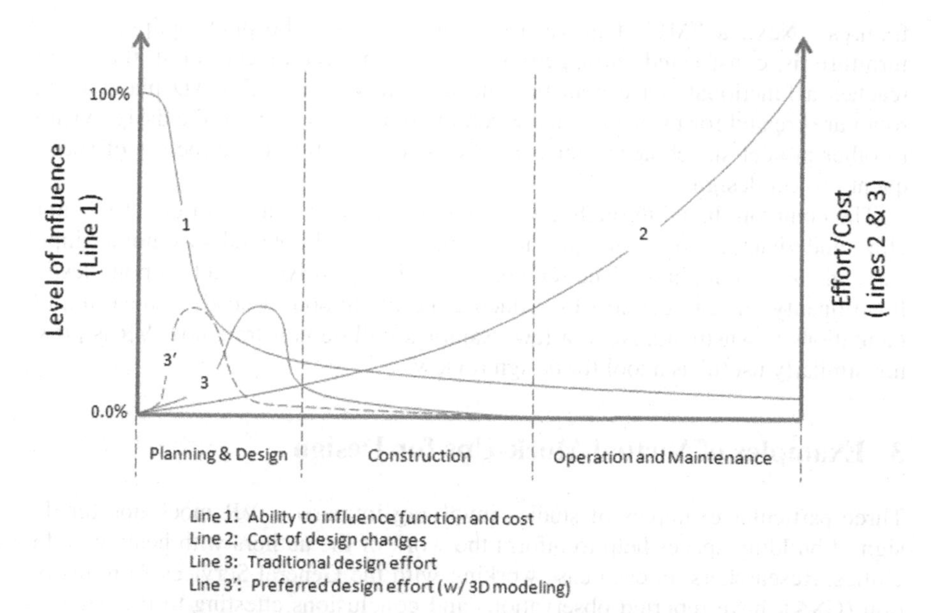


**Fig. 1.** Chart illustrating the relative value of being able to settle design decisions sooner in the project life cycle (adapted from Eastman et al., 2008)

identify and implement design changes easily and inexpensively. The value to be derived from the use of mock-ups of some type during the design phase is underscored by the experiences of stakeholders who have employed specialized consultants who provide early prototyping services. Notable industry examples are the Rapid Prototyping Mock-Ups employed by design consultant IDEO and the Hill-Rom RoomBuilder™ Workshops.

Hamilton and Himwich (2008) highlighted IDEO's use of rapid prototyping with cardboard, at the schematic design stage, to review the functional aspects of building designs for the Presbyterian Healthcare Services of New Mexico. Routine and emergency clinical scenarios, involving 40 different interdisciplinary teams, were staged over a course of two weeks to test how well requirements were met in the proposed design. Critical design deficiencies were addressed in no less than seven different units of the hospital. Modifications included spatial relationships and system architecture such as the waste disposal systems and the orientation of equipment with respect to surgical processes. The final design was greatly enhanced by stakeholders' detailed input.

At their Customer Experience Center in Batesville, Indiana, USA, Hill-Rom, Inc. offers their RoomBuilder™ Workshop, a three-day design activity intended to bring customers into the design process early and help them arrive at final schematic designs that maximize the functionality of specific hospital rooms. Typically starting from the 2D floor plan brought by the customer (owner and maybe architect), facilitators guide the workshop team through experimenting with room layouts using 2D tabletop kits that include scaled walls, equipment, hardware, and fixtures. Next, a PMU of moveable foam-core walls, hospital equipment, and furniture is constructed, arranged, reviewed, and rearranged until the group reaches a functional space plan that meets their vision. 3D CAD plans of the room are created for inspection and given to attendees to present the design vision to other project stakeholders, thus significantly impacting the direction of subsequent project design.

The common thread through these examples is the utilization of effective modeling and visualization techniques to obtain broad and detailed stakeholder input early in the design phase. The reported result is a positive impact on both design functionality and project cost by reducing the likelihood of costly change implementations downstream. Next, a few examples will demonstrate how VR is proving similarly useful as a tool for design review.

## 3 Examples of Virtual Mock-Ups for Design

Three particular examples of studies involving immersive VR mock-ups for design of building spaces help to inform the work of the authors with healthcare facilities. Researchers, in each case working with the General Services Administration (GSA), have reported observations and conclusions attesting to the value of immersive VR as an effective 3D visualization medium to appropriate for design.

Chan (2005) at Iowa State described experience gained over three years in developing a model of the GSA's Adaptable Workspace Laboratory (AWL) for display in a six-sided CAVE™ display. Chan sought to explore the use of virtual environments for analyzing interior spaces with respect to factors such as employee productivity. Periodic review and recommendations from the American Institute of Architects (AIA) and GSA partner participants aided in improving accuracy and reducing rendering complexity. Aesthetic inaccuracies in lighting, material textures, color schemes, and sound quality were also addressed. Movable furniture and personalized items like notepaper were added to increase utility and realism. Acknowledging a huge time investment to create high-fidelity AWL model rendering, Chan concluded that design firms should explore the use of virtual environments.

Majumdar et al. (2006) sought to test whether VR mock-ups can improve the design review process. They used the Walt Disney Imagineering Computer Assisted Virtual Environment (CAVE) facility (to be distinguished from the trademarked CAVE Automatic Virtual Environment (CAVE™)) to display and test a VR mock-up a federal courtroom design with a low level of detail, excluding such aspects as room aesthetics and lighting. After first convening a review meeting with stakeholders other than the federal judges to identify needed changes and correctable errors, the researchers let judges (key end-users) review the courtroom design and make suggestions for final designs. The review process was shortened from the typical eight hours to three hours. The researchers noted increased capability to make timely modifications, enhanced focus of the collective attention of the group of reviewers on each issue, and ease of communication between the owners, architects, and model builders via 3D CAD drawings developed for the design review sessions.

In the final example illustrating the value of VR mock-ups, Maldovan et al. (2006) reported benefits realized by utilizing a VR mock-up for design review. Their objectives were to identify and incorporate attributes of a typical design review meeting into a VR mock-up review meeting and to improve the process for analyzing designs. Also working with a VR mock-up of a federal courtroom, the researchers engaged representatives from the federal courts, the design firm, and the GSA in two separate meetings. The first meeting involved non-project specific stakeholders to identify perceived deficiencies and to rank and prioritize tasks for a successful courtroom design review. This meeting produced review objectives for the second group. After the model was updated to reflect needs identified by the first group, the project-specific group was enlisted to rank the model according to the established objectives. Categories included sightlines, aesthetics, lighting, security, and ergonomics. The VR mock-up was deemed beneficial to the design review process because the main review tasks were correctly accomplished for less cost and less time compared to utilizing a PMU.

## 4 Purdue Patient Room Mock-Up

In light of these strong motivations for exploring the utility of VR mock-ups for healthcare facility design review, the authors developed an exact virtual replica of an existing patient room and furnishings in the Bariatrics and Obstetrics Department at St. Vincent Indianapolis Hospital. Maintained at the Envision Center for Data Perceptualization at Purdue University, this patient room VR mock-up is set up to run in the Center's CAVE-like setup, a Fakespace FLEX™ VR theatre system featuring three 3-meter by 2.4-meter panels for active stereo rear projection. A user dons special eyeware for 3D perception of the projected and uses a tracked handheld device to interact with the virtual model at a true 1-to-1 scale. An audio recording of hallway noise obtained from St. Vincent is played and changes volume as the user moves to different locations in the room or opens and closes doors. Open source toolkits allow for flexibility and portability of the VR mock-up. Further details of the hardware and software are reported by Dunston et al. (2007).

### 4.1 Real-Time Interactive Elements

In addition to the sense of presence derived from the immersive display, the real advantage of this VR mock-up over other 3D modeling media is the interactivity, which includes both changes in viewing perspective corresponding to movement through the model scene and the ability to handle virtual objects in the scene. The VR mock-up contains numerous furniture and equipment objects which can be rearranged via the handheld device. In addition to changing the positions of these items in the room, the user can also experience a variety of lighting levels. The size of the room also can be changed between two preset dimensions by moving the wall opposite the headwall.

These realistic interactive features are the critical value-adding aspects derived from VR mock-ups because they enable healthcare practitioner users to evaluate

**Fig. 2.** Views related to doors: (a) initial view from outside the patient room and (b) recognition of unused space and clearance between open entry and bathroom doors.

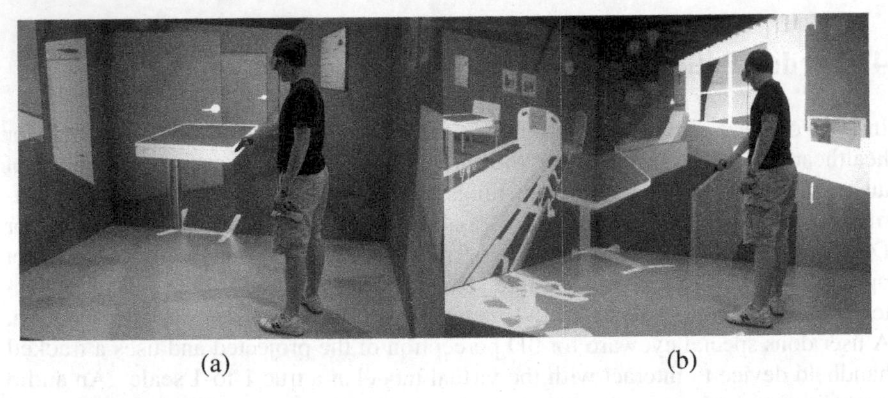

**Fig. 3.** Interaction examples (a) checking room for moving a bedside tray table and (b) observing clearance for opening a cabinet

more than just a static layout but the actual dynamics of the environment in which care will be provided to the patient. Adequacy of space, features posing safety risk, and specific viewing perspectives can all be inspected and assessed to evaluate the overall functionality of the design via this kind of VR mock-up. Several examples are highlighted below to underscore the usefulness of the VR visualization.

Figure 2 illustrates two particular aspects of the patient room design that may be uniquely evaluated in the immersive environment. Figure 2a shows the approaching view of the room to a visitor if the door is open, which depicts the point at which visitor and patient make their first visual connection and impression. Figure 2b illustrates the fact that in this design, the entry door and the bathroom door enclose a corner of the room when both are open which creates an inaccessible space under this condition. The clearance between the two doors as they swing open can also be inspected.

Figure 3 illustrates some of the VR mock-up's capacity for interaction. The user in Figure 3a is moving the hospital bed tray table, and the same user in

Figure 3b is able to open the doors of a floor cabinet and check the clearance that is available when the tray table is also on that side of the bed. The space is clearly seen to be very tight. Many such possibilities for reconfiguring and inspecting the furniture and mobile equipment arrangements are available to the user during the review session.

## 4.2 Viewing the Design from Unique Perspectives

Figure 4 provides an example of a how the VR mock-up can be used to present the reviewer with a unique perspective on the design of the patient room and also compares the desktop and immersive display views. Figure 4a depicts a user examining the VR mock-up from the perspective of lying on the hospital bed. Items of interest from this perspective might include the view of the entry door, the view of scenes outside the window, the view of the television and wall clock, the view obstructed by the footboard of the hospital bed, or the placement of overhead lighting.

One might consider whether these types of views cannot also be satisfactorily inspected in a 3D model displayed on a conventional monitor. Figure 4b shows the desktop view corresponding to the immersive view in Figure 4a. While many of the same details are visible, the immersive quality of the CAVE-like display provides a better sense of spatial awareness as well as a wider field of display.

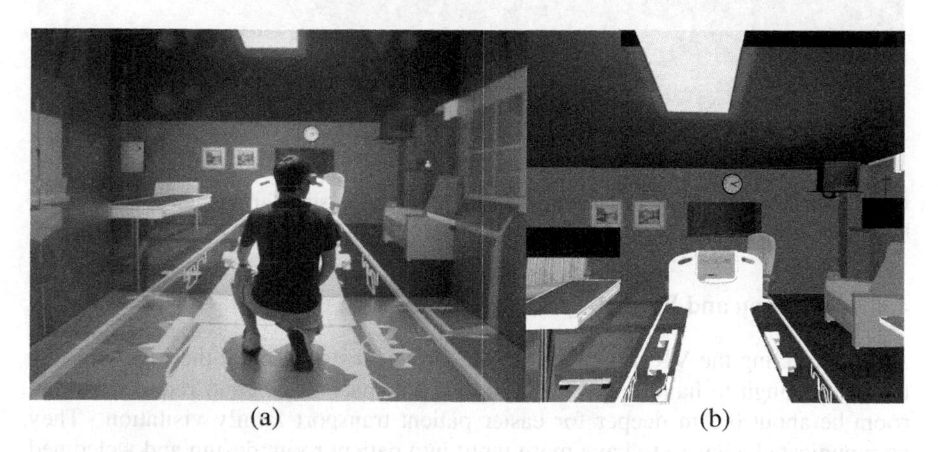

(a) (b)

**Fig. 4.** Comparison of the patient's view from the bed between (a) immersive VR and the (b) desktop VR platforms

## 4.3 Identifying Safety Hazards

Figure 5 shows an example of how a design feature can be identified in the VR mock-up review that has safety implications for both the patient and the care provider. As the individual in the Figure 5a scene stands in front of the sink, he recognizes that the shelving to the left above the sink extends far enough over the sink to present a strike hazard for someone bending over to use the sink. If such an instance is identified early, the hazard can be eliminated by either modifying the shelf design or specifying a different cabinet-sink design. In this instance, it is arguable that the desktop monitor display of this same area as depicted in Figure 5b might also be adequate to recognize the strike hazard because of the small area involved. However, the lack of presence in the scene leaves the viewer without the enriched understanding of space as revealed when one's own body is a part of the viewed scene, i.e., being *at the sink* provides a better understanding of spatial implications for human mobility in the designed space. Of course, many such design implications as this would be overlooked if the design and room layout were presented to stakeholders only in the form of a 2D plan. Even in a 3D walkthrough, the lack of immersion in the scene may still result in the reviewer missing these kinds of spatial shortcomings in the design.

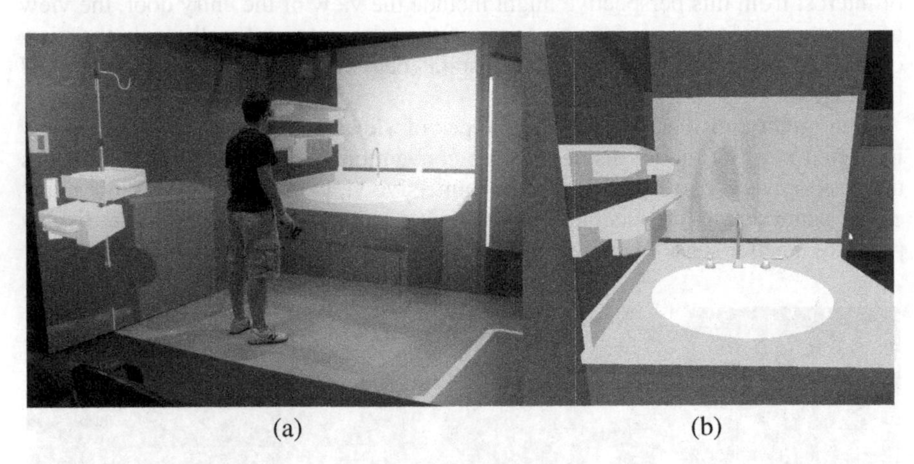

(a) (b)

**Fig. 5.** Comparable views of shelving extending over the sink far enough to be a strike hazard as seen by (a) immersive VR and (b) desktop VR platforms

## 4.4 Evaluation and Potential Enhancements

After inspecting the VR mock-up, St. Vincent nurses stated that the mock-up was realistic enough to have helped them anticipate space issues and request that the room be about 0.6 m deeper for easier patient transport family visitation. They communicated a desire to have more input into patient room design and welcomed this method for facilitating it.

Many spatial enhancements to the existing simulation are possible. With the review occurring earlier in the design process, users could be given more control over structural changes via manipulating the size and shape of the room. The ability to shift (without presets), add, or remove walls or ceilings would allow users greater freedom to explore alternative room configurations. Virtual humans (also known as avatars) could be programmed to simulate typical tasks performed by care providers, providing an additional cue for scale and size issues and demonstrating how multiple occupants crowd the space. Specialized equipment such as ceiling lifts could be added to illustrate function within the space and either could be operated by the users or demonstrated by the avatars. Because occupants may be many different heights, the size and scale of the avatars or the room itself could be controlled by users to better understand how the room would suit a variety of different patients and providers.

New visual effects can be added. In addition to the room's existing lighting simulation, external lighting could be added to simulate various times of day, as well as other concerns such as external lights aimed into a patient room after dark. Other external elements such as surrounding buildings and landscapes could be created, and the view from the patient room could be changed to simulate the view from any room in the facility.

Although the current simulation includes basic sound capabilities, more sophisticated aural enhancements are possible. Different types of wall coverings and other construction materials have distinct effects on sound, potentially leading to a healthier environment for patients. These material effects can be modeled and used to demonstrate the acoustic implications of alternative construction materials.

The programming required for an interactive VR mock-up is time-consuming, and adding these enhancements will multiply that time requirement. Measures, therefore, must be taken to shorten model development time and cost. This objective should be accomplished also for other hospital units such as operating rooms, nurse work stations, neonatal intensive care units, intensive care units, emergency rooms, etc. The developers of the Purdue patient room VR mock-up are presently pursuing these objectives (Dunston et al., 2007).

## 5 Conclusion

While advanced industry practices have demonstrated the value of mock-ups for enhancing the impact of design reviews in the early stages of design for interior building spaces, research has demonstrated the technical efficacy of immersive VR mock-ups for this purpose. Experience with the Purdue patient room VR mock-up demonstrates the advantages of an interactive immersive environment for hospital design review. It is reasonable to expect that space and equipment functionality may be more extensively validated through added interaction capabilities. Further developments and studies are required to confirm the wider range of application and the cost effectiveness of broadly adopting VR mock-ups into hospital design practice.

## References

Chan, C.S.: Visualizing the Effectiveness of Work Space in a Virtual Environment, AIA / GSA Project Final Report, Iowa State Univ., Ames, IA (2005)

Dunston, P.S., Arns, L.L., McGlothlin, J.D.: An immersive virtual reality mock-up for design review of hospital patient rooms. In: Messner, J.I., Žikić, N. (eds.) 7th International Conference on Construction Applications of Virtual Reality (CONVR 2007), Penn State University, pp. 139–146 (2007)

Eastman, C., Teicholz, P., Sacks, R., Liston, K.: BIM Handbook: A Guide to Building Information Modeling for Owners, Managers, Designers, Engineers, and Contractors. John Wiley and Sons, Hoboken (2008)

Hamilton, B., Himwich, D.B.: The smart choice: test drive your building plan through rapid prototype mock-ups. Healthcare Design 8(9), 26–29 (2008)

Hamilton, D.K.: The four levels of evidence-based practice. Healthcare Design 3(4), 18–26 (2003)

Hamilton, D.K., Watkins, D.H.: Evidence-Based Design for Multiple Building Types. John Wiley & Sons, Hoboken (2009)

Majumdar, T., Fischer, M.A., Schwegler, B.R.: Conceptual design review with a virtual reality mock-up model. In: Rivard, H., Miresco, E., Melham, H. (eds.) Building on IT: Joint International Conference on Computing and Decision Making in Civil and Building Engineering, Montreal, Canada, June 14-16, pp. 2902–2911 (2006)

Maldovan, K.D., Messner, J.I., Faddoul, M.: Framework for reviewing mock-ups in an immersive environment. In: Raymond Issa, R. (ed.) CONVR 2006: 6th International Conference on Construction Applications of Virtual Reality, Orlando, Florida, August 3-4 (2006), available on CD, 6 pages

# The Immersive Virtual Environment Design Studio

Marc Aurel Schnabel

Chinese University of Hong Kong, Hong Kong

**Abstract.** This chapter discusses the implications of designing, perception, comprehension, communication and collaboration within the framework of an 'Immersive Virtual Environment Design Studio'. It has been suggested that the unique properties of immersive virtual environments can empower designers to express, explore and convey their imagination more easily. For these reasons the very different nature of virtual environments may allow architects to create novel designs that make use of additional properties that other realms cannot offer. An architectural design studio was conducted to investigate the relative effectiveness of immersive virtual environment as environment for creation, interpretation, communication of spatial design, and collaboration within a design team. The outcomes of the design studio present how immersive virtual environments successfully aid architectural designing.

**Keywords:** architectural design, collaboration, design medium, immersive virtual environments.

## 1 Introduction

There is a distance between the imagination of a design and its representation, communication and realisation: architects use a variety of tools to bridge this gap (Pérez-Gómez and Pelletier, 1997).

Designing in Virtual Environments (VE) may minimise this divergence. As a result of this designers are empowered to express, explore and convey their imagination with fewer differences (Dobson, 1998). Most researches on VE within the architectural context have focused on their use as presentation or simulation environments. Only some research is done that studies the use of VE for designing (Gero, 1999a).

Design is an activity that is greatly complex and influenced by numerous factors. It has been suggested that for this reason, the very different nature of VE may allow architects to create designs that differ from not VE aided design (Davidson and Campbell, 1996). This chapter examines the implications on architectural design by using immersive computer generated VE as the medium for design in a collaborative setting.

X. Wang & J.J.-H. Tsai (Eds.): Collaborative Design in Virtual Environments, ISCA 48, pp. 177–191.
springerlink.com 

## 2 Related Research

Architectural design within VE has been widely used as a method of design simulation and presentation. Educational and professional settings employ VE successfully to study, communicate and present architectural design. The rapid development of digital tools during the past decades had profound impacts on the architectural education and the way how architects create, converse or appreciate 3D spatial environments (Koutamanis, 1999). Numerous publications illustrate the impact digital media had on architectural design and propose solutions for multi-media design studios and how to make use of VE (Maver, 2002). Dave (1995) investigated distributed design studio. Wenz and Hirschberg (1997) studied collaborative design within remote collaboration, while Hirschberg et al. (1999) analyzed pattern of communication within digital design studios. VE often became a presentation tool only to assess design alternatives and final design solutions (Achten et al., 1999). Yet, they did not look into the comprehension and conception of design within Immersive VE (IVE).

One particular form of design studio emerged in the early nineties that investigated various possibilities digital media and VE can offer to the learning and the exploring of architectural design. These so called *Virtual Design Studios* (VDS) defined virtuality as acting while physically distant or as acting by employing digital tools (Maher et al., 2000). Yet virtual did not refer to an IVE. Instead, VE were established by the choice of design (Achten, 2001), way of communication (Schmitt, 1997) or tools (Kurmann, 1995; Regenbrecht et al., 2000; de Vries and Achten, 2002). Yet a significant potential of research remained unexplored. Within the context of a VDS, a real immersion into a VE could be used for designing.

## 3 Architectural Design in Virtual Environments

The question of describing architectural design itself is central to any debate about the use of VE to support architectural design. There are many definitions of design. Lawson (1980, 1994) for example defines architectural design as an act of creation, while Dorst (1996) tries to address more general definition of design related to any action within an artistic intention. Others have considered the difference between design in physical worlds and design in VE (Bridges and Charitos, 1997).

Obviously, there are significant differences between manual drawings, sketches, paintings, etc. and images and models produced using CAD systems. Since the introduction of CAAD by Sutherland's Sketchpad (Sutherland, 1963), extensive research has been undertaken to explore various possibilities and potentials of different media and realms (Goldschmidt, 1991; Schön and Wiggins, 1992; Robbins, 1994; Goel, 1995; Lawson and Loke, 1997; Suwa and Tversky, 1997; Verstijnen et al., 1998; Schnabel et al., 2007).

Nevertheless in all descriptions of the design development, we find in common the activities of cogitation, expression/modelling and communication/testing (Broadbent, 1973). Without engaging in discussion about computer-mediated cogitation or speculating on what the results of VE supported design may be, it is obvious that the tool used is affecting the process and outcome of a design. The impact of the medium of representation on the content has been the subject of many studies. Marshall McLuhan (1964) proclaimed that 'the medium is the message'. In the same way 2D drawings had a significant impact on architectural design in the 15th century, hence it is logical to assume that digital media, especially IVE, have a considerable impact on architects' ability to conceive, understand and communicate spatial environments.

### 3.1 Initial Design Stages

VE play an increasing role in architectural design especially during early design activities (Bertol and Foell, 1997). Equipment and software to engage in VE are easily available and particularly affordable. However, not sufficient attention has been paid to the results and possibilities of architectural design in IVE (Stuart, 1996). Lessons learnt from academic contexts have been employed in various commercial settings within the creative industries. A few corporations and architectural companies make use of global locations, sharing of resources and work force. This collaboration goes beyond video-conferencing or file-sharing and includes shared design sessions and expertise consultation (Burry et al., 2001).

The overall dimensions of an architect's final 'product' as well as constraints of resources make it usually inevitable that architects communicate and express their intentions with the help of models. Architects can make use of real and virtual instruments to translate and communicate their designs in these mixed realms (Schnabel, 2009a). In recent years, computer generated VE are increasingly used as a device of communication and presentation of design intentions (Leach, 2002). VE are employed successfully to study, communicate and present architectural design (Bertol and Foell, 1997). However, according to Maze (2002), IVE are seldom used in initial design stages for creation, development, form finding and collaboration of architectural design.

### 3.2 Computer Supported Collaborative Design - Virtual Design Studio

As Kvan (1999) argues, Computer Supported Collaborative Design (CSCD) can enhance the exploration of ideas and their communication. Kalay (1998) describes the difficult situation of architectural ventures that employ CSCD methods. Integrated projects are undertaken by fragmented teams, leading to decreased performance of both processes and products. Virtual Design Studios (VDS, 1993-2002) have been widely used in the last decade as an environment for architectural design teaching. Immersion has not been used for design interaction, although shared immersive virtual spaces have been employed for design reviews (Davidson and Campbell, 1996). The next logical step to develop the VDS was, therefore, to establish joint design sessions where users can collaboratively create, interpret and communicate design ideas within an IVE and to examine if this context offers any new opportunities or solutions to problems encountered. Mitchell (1995) argued that there is indeed a need for an ongoing evolution of the

VDS towards a fully integrated studio where the borderlines between realms are dismantled.

# 4 Designing in Virtual Environments

The objective of our research was to identify how designers perceive architectural space within VE by looking at the creation, interpretation and communication of architectural design in a collaborative setting. In order to investigate the relative effectiveness of IVE, we conducted a series of experiments. With this, we want to understand if form comprehension and form finding is enhanced within VE activity (Schnabel, 2004).

Therefore, we set up an experiment, a design studio that enabled students to design within a VE that imbeds immersive tools into a broad context of CSCD. The experiment engaged students in typical architectural design contexts. We were then able to draw conclusions that are relevant to the praxis of the architectural design profession. The Studio Experiment, called 'Virtual Environment Design Studio' (VeDS), look into conditions of architectural design, collaboration, communication, understanding and re-representation within a setting of an IVE design studio. It simulates a normal design process within the architectural profession by its nature (Schnabel and Kvan, 2003).

We explored factors influencing designers during the design process and investigated the relationship of 3D space within VE versus the physical realm.

As already mentioned above, instruments and designers have an influence on the outcome. Hence we engaged students of both genders of the master's programmes at the departments of architecture of the University of Hong Kong (HKU) and the Bauhaus University Weimar (BUW), Germany.

# 5 Immersive Virtual Environment Design Studio (VeDS)

The design studio is the established context for architectural learning. Collaborative learning and designing have been demonstrated to support effective learning in architectural design (Kvan et al., 1999). As a result, we explored in more detail how a VDS, that is truly virtual and designers are immersed into a VE, affect the process and outcome of design (Schnabel, 2002).

We sought tasks that engaged designers at different levels of complexity within VE. Thus, we decided upon the design of a commercial helicopter landing station in the urban setting of Hong Kong. This task required students to work in three dimensions at all times and to fully navigate a 3D (not 2.5D) space. The IVE interface did intentionally not to allow a detailed modelling and users could therefore only establish initial layout of solid and voids in order to generate their design proposals. In this assignment, the designer could work in a virtual model of Hong Kong from the viewpoint of the pilot flying to and from the site or of the passenger waiting at the helipad to embark.

The assignment for teams was split in two parts, one for each team: either the land- or the airside of the helipad. Additionally each part of the task had one static and formal, as well as one dynamic and path focused. Both of which had to be addressed in the design proposal (Table 1). The teams gained authority of a design area and at the same time had to negotiate with the remote team, who worked on another area of the same site. Both teams had to collaborate in order to reach a solution while they also had enough freedom to explore their own design aspirations.

**Table 1.** Design task distribution: Landside/Airside and Static/Dynamic

| Team | Authority | Programme | Component |
|---|---|---|---|
| *A* | *Landside* | *Check-in/Waiting enclosure for passengers* | → *static* |
| | | *Driveway/Parking* | → *dynamic* |
| *B* | *Airside* | *Control tower for Air controllers and tourists* | → *static* |
| | | *Apron/Flying* | → *dynamic* |

We modified the *Virtual Reality Architectural Modeller* (VRAM) software by Regenbrecht et al. (2000), and added new input features based on gestures, called *VRAM/G (Seichter, 2001).* Comparable to the input for handheld computers or PDA devices, users gestured with the *Stylus* and their movements were tracked by the tracking device and, via one computer, translated by VRAM/G into basic 3D primitives. Table 2 shows the reference guide of gestures that the software recognises and translates into relevant primitives. A set of object libraries allows variations of a set of primitives.

**Table 2.** Gesture Reference Guide

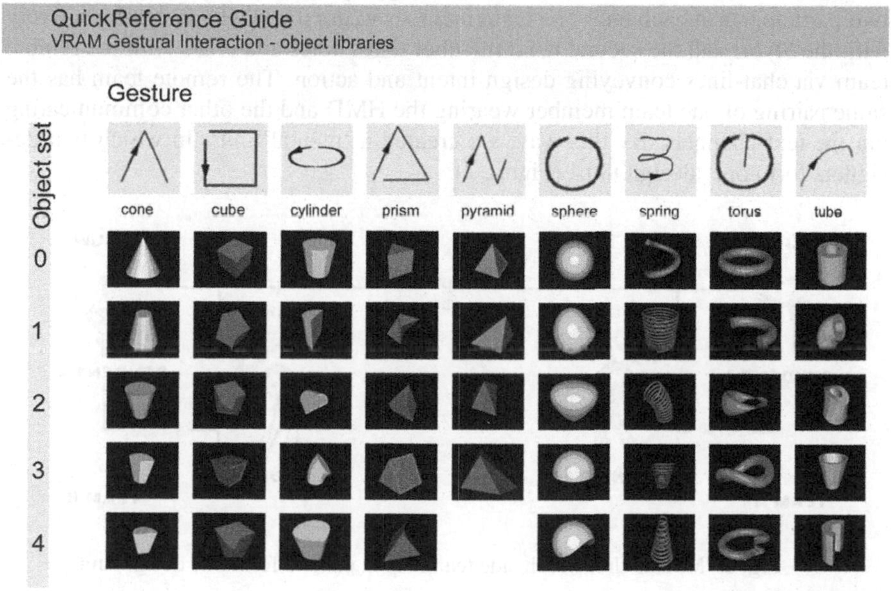

QuickReference Guide
VRAM Gestural Interaction - object libraries

Gesture

Object set

cone cube cylinder prism pyramid sphere spring torus tube

0
1
2
3
4

A second PC was used as communication channel, using *ICQ*-software, an internet browser *(Internet Explorer)*, a web-based database (Figure 1) as well as other presentation software *(AutoDesk 3DStudio VIZ* and *Adobe Photoshop)*.

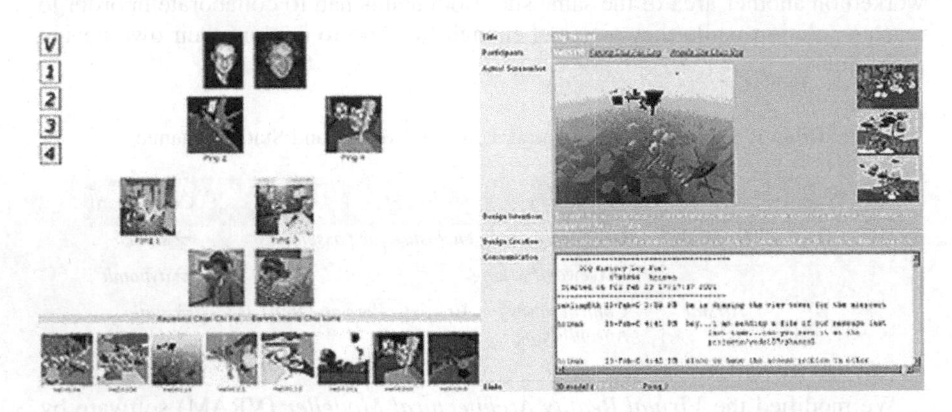

**Fig. 1.** Screenshots of Database of VeDS; Left: Overview of output by one of the teams; Right: Presentation of work in one phase with text annotation {http://courses.arch.hku.hk/vds/veds01/db}

As in a moderated discussion session where the microphone is passed to speakers, the *Stylus* was virtually passed between the teams on each remote side and the resultant design sketches were produced within the IVE in the course of the alternating sessions. To support the design process more fully, text communication was also provided (Wong and Kvan, 1999). In order to capture the design intent, we used a modified 'think aloud' methodology by establishing a design team of two participants at each end: one team member wears the HMD designing actively with the *Stylus* and the second team member takes notes and chats with the remote team via chat-lines conveying design intent and action. The remote team has the same pairing of one team member wearing the HMD and the other communicating via the text-channels. By this way, we created a 'mental unit', in which two designers form one 'design unit' (Figure 2).

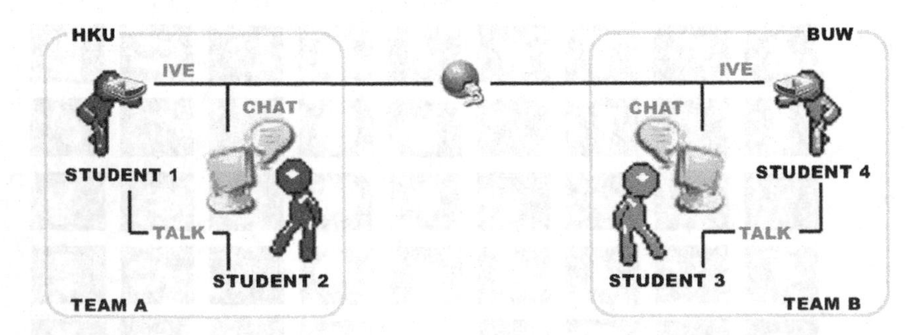


**Fig. 2.** 'Mental Unit': each side teams up in pairs to form one design unit

The designer using the HMD was free of all complex communication actions that could hinder the smooth flow of the act of design. While the other team member was only writing and corresponding with the remote team, a record of all communication was automatically generated. In this way, the text records provided a protocol to be analysed.

Although in the past our goal has been to engage in heterogeneous environments, with each participant using whatever equipment they wish, the problems of VE collaboration precluded such freedom. In this experiment, both universities employed the same configurations and used the same immersive VR equipment and all participants of these experiments received thorough instructions on the use of the equipment prior to the studio.

The design tasks were specified in order to present the students with assignments that are appropriate in scale, content and effort to the medium available. Special care was taken neither to favour nor to hinder the designers in creativity and translation of idea and result.

# 6 Discussion

This experiment showed that students using IVE engage actively in exploring space and volume in an inclusive manner. We validated that it is possible to successfully design, communicate and collaborate in IVE using an architectural design studio setting. Students communicated their spatial ideas using 3D objects that are accessible and interpretable by others in close relationship with the original design intent. This can either be very lively and similar to a sketch or as complex as a full design studio. Although it was possible that the teams would concede to the technical complexity of the system and the difficulty of working together, the teams did engage in collaborative work, building up, step by step, on the work of the efforts of team partners and preceding steps.

All teams have succeeded in designing a helipad and thus completed their tasks. The VeDS allowed a variety of design outcomes since every design studio has its own characteristic and every design-team their own approaches. Consequently, the results can be analysed in various ways. We explored the essence of the progress, form and communication of the students' works as described below. General speaking, the outcomes of the VeDS confirm that collaboration and designing within IVE lead to successful and valid architectural design schemes (Schnabel and Kvan, 2002).

## 6.1 Progress and Form

A review of the graphic results and digital models shows that students used the 3D design space actively. Volumes were created to represent design elements at all cases within the available 3D design space. Typically, a design created in a 2D space would have placed elements in plan with some raised in section/elevation to create 3D spaces. In the experiment, however, the students started 'drawing' the design elements at all points above the ground plane. Observation during the creation processes of the designs showed that participants did not use a 'bottom-to-top' (floor by floor), an 'inside-out' (function defines form) or 'outside-in' (form defines function) approach to their design. Mostly, students used an integrated design method by making use of all approaches almost at the same time. Being virtually inside the model, they sculpted their proposals, employing the flexibility of viewpoints offered in IVE. They explored the spatial impact of their design proposals in relation to existing forms and activities from outside and inside the model. They changed their viewpoint constantly from a general macro- or overview to a micro view of details in order to check spatial relationships and design proportion of the various elements of their design.

Although the input systems were crude and clumsy, users rapidly learnt to represent their design intent by using the available representational volumes: cubes, cylinders, cones and spheres. The objective of the assignment was to establish initial relationships of form, space, solid and void, in order to establish the overall gestalt of the design, while a detailed planning was not required.

The software offered a variety of library elements. Nevertheless, the teams predominantly made only use of the most basic set of library elements. Despite this constraint, these primitives offered students a significant range of design expression. This matches the simplicity of the software, its interface and its operation that did not require any complex menus or operational overheads or special setups. Similar to conventional design media, such as paper and pen, initial ideas were sketched freely into the space. Equivalent to a 2D medium, the various shapes symbolised both positive and negative representations of form and space or symbolised just reference points or other graphical elements. Viewers of the models, however, were able to understand this ambiguity of this 3D sketching. Yet the three-dimensionality of the primitives and the ability to change viewpoints and scale allowed the students to explore the space in a way that a sketch cannot offer.

In some cases, because of the lack of experience, problems with the hard- or software or the complexity of IVE, errors occurred, but often they were transformed into meaningful solutions, a design behaviour which is observed in other traditional 2D design environments as well (Schön and Wiggins, 1992). For example, instead of deleting unwanted elements, some teams chose to keep those elements and integrate them into their design. These 'errors' were then actively repeated to generate the desired outcome.

Each VDS merges the diverse backgrounds and skills of participants, their education, knowledge and understanding of architectural design. Subsequently these differences are expressed in the design. Since each team had authority over its own area, the VeDS allowed a distinct development of each team's design and if desired by the teams a clear distinction between the two design areas. In the course of the studio, it was necessary for students to agree on common readings as well as shared definitions of each team's working space, their borders and common elements and parts as well as working and design strategies.

Participants noted in the chat line communications that the resultant designs surprised them in their ingenuity and presentation. Constrained to their own preconceptions, they realised that working with the tool within IVE was not only easier than initially anticipated, but also have created outcomes that were superior to those their own skills enabled them within other design media. It appears that IVE allowed students to experience their ideas in ways that are different from non-immersive environments. They reported that the interaction of idea and creation was direct, that each stroke had an immediate impact on the design. They could not only understand easily the intentions of the other teams but also were inspired by the ease and freedom the tool and the environment offered them. For the students, it seemed that they communicated directly with their model and being part of it, instead of being an only distant designer. The students stated that this kind of design method has led to new forms and new arrangements. Nevertheless, for all students it is a novelty to design within IVE. Architectural design studios have not developed much further from conventional design methods despite the available technologies and innovative topics. Students are therefore not used to designing three-dimensionally and interactively, as it is possible with IVE.

### 6.2 Communication

The experiments proved that the teams communicated confidently as anticipated, by means of the set up, of which two participants formed one communication unit to correspond with the remote team. The teams intensively discussed issues of design, concepts and form. Due to the nature of the task and application, the groups had to formulate their intentions and discuss them with their remote partners in order to develop their scheme further. In addition, participants developed a personal interest in sharing their experience and creation with their colleagues and other teams.

In the analysis of the chat protocols showed only a few lines of navigation/orientation discussions. This suggests that participants could not only orientate themselves easily within VE, but they were also able to abstract and extract the design intent of the remote partner without much difficulty. Neither the tool nor the environment was an issue to talk about because both of them have blended into the design process harmoniously.

While the text records do not identify how or why the students were using the 3D space in these ways, we do find records of intense discussion about design, functions and concepts.

By referring to the images they saw in the model provided by their distant collaborators, students could engage in design discussion and development of the scheme. VE did change how the students developed and expressed their ideas. This new way was then communicated to the remote team using the design itself and discussion it on chat texts.

We noticed that participants from BUW tended to deal with more conceptual schemas while HKU students tended to be more factual, specific and they described ideas in tangible terms, possibly reflecting the educational characteristics of the two institutions. With such distinctions, it is notable that the VR environment supported these differences and the collaboration was successful.

## 7 Conclusion

We developed our experiments based on reported results of prior research in design collaboration and communication using real and virtual environments

(Schmitt, 1998). We carried out an architectural virtual design studio that took the issues of IVE into a complete architectural design scenario. Then we transferred our experiences of the VeDS to some experiments with abstract problem solving.

> *In a similar context Dave (1995) also confirms that VE is a constructive tool to support the design and communication process at least in establishing co-presence for a shared experience in spatial review. Yet how is this support extended to a design setting? Chat-protocols show participants remarking to each other that the collaborative experience was satisfying. That means, in IVE the exploration of space, volume and location is enhanced and site-specific problems are not only better recognised, but also possibilities are investigated. This is an improvement over other forms of design sharing that is analogous to the conclusion drawn by Campbell and Wells (1994).*

Using a 2D medium to translate spatial ideas apparently reduces the exploration and communication of volume and space. Coherent to our findings, Dorta (2001) concludes that VE have significant impact on the activities of communicating 3D information within a design process caused by the impact of VE on the cognitive aspects of the design activity: the formation of 3D mental images, visual perception and mental work load. The results of the experiment support those findings. VE permit an enhanced understanding of spatial compositions. In other words, using VE as medium to design spatial 3D compositions, designers can pursue ideas with a smaller cognitive workload.

While users of VE can change their viewpoints and escape gravity, they also maintain the feeling of presence within the digital 3D models are generated with the intention of conveying overall design intentions similar to physical models, constructed to improve the perception of designs developed by drawings. As a result, IVE provide an immediate feedback to users that are not possible within CAD or traditional design media (Chen, 1995). It appears, therefore, that designers can work more three-dimensionally within VE since every object is experienced through movement and interaction. The design is created as a whole entity within space and not as a 2D planar representation. This possibility offers a different 'conversation' with the design that otherwise is not obvious or possible. In addition, spatial issues can be addressed in a manner akin to the real world. The design process becomes more immediate, in some aspects, with the tools available enhancing the translation of the designers' and users' mental intention into spatial objects and 3D design decisions. Subsequently these possibilities have an impact on the quality of the resulting design. The experiences seem valuable even in spite of the amount of technological overhead used and the abstract realism of VE.

Only in very recent years architectural design is evolving beyond the traditional language (Gruber et al., 2003). Architects discover new ways and different tools to communicate their design (Schnabel et al., 2007). Hereby VE can help them to explore and express ideas unlike traditional methods.

A similar phenomenon happens within the academic and educational environment. Less than a decade ago many schools of architecture did not allow students to deliver CAD drawings for design projects assuming that those limit the exploration and understanding of design. In fact, the early experiments in using the computer in the design process quite often failed only because of the restrictions of the available hard- and software. Today, students are familiar with CAD software even before they enter the university (Dokonal and Hirschberg, 2003). Still many questions remain unanswered and new questions arise in the relationship between architectural design and digital media. Architectural design is both an imagination and the ability to convey this idea. The teaching of architectural design has now the possibilities to make use of the advantages that VE can offer without loosing the qualities of the established conventional methods. Yet too often however, in using digital media and VE tools the students are more conversant than the teachers are. All those changed the dynamics of architectural education. Yet, this has to be reflected in how we teach.

Finally we recognised that the translation of design from VE into other media is potentially problematic (Gero, 1999b), suggesting that developments may be needed to facilitate the making of physical models. Similar to Gibson and Kvan's (2002) findings this suggests that other technologies such as rapid prototyping or automated construction methods may have a significant contribution to make to a design process that engages VE. Synergies between the different realms, media and technologies can be developed in a collaborative environment that fosters the evolution of new kinds of forms and structures.

This introduces a new way of designing and therefore fits well within existing paradigms (Mitchell, 1994). Wiener (1954) predicts the future merging of location and culture. Referring how physical and virtual architecture is an expression of cultural understanding. Both have their own properties, but both deal with the same matter. This will result in new understanding of architectural design and vice versa, this understanding influences the definition of architectural design (Schnabel, 2009b).

The potentials of VE are obvious and omni-present, yet they are not explored fully to their own capacities. As Maver (1973) postulates: "Design follows its own paradigms". Therefore, it evolves and re-establishes itself by its own developed expression.

## References

Achten, H.H.: Normative positions in architectural design - deriving and applying design methods. In: Penttilä, H. (ed.) Architectural Information Management - 19th Conference on Education in Computer Aided Architectural Design in Europe, eCAADe, Helsinki, Finland, pp. 263–268 (2001)

Achten, H., Roelen, W., Boekholt, J. T., Turksma, A., Jessurun, J.: Virtual reality in the design studio: The eindhoven perspective. In: Brown, A., Knight, M.B.P. (eds.) Architectural Computing from Turing to 2000 - 17th Conference on Education in Computer Aided Architectural Design in Europe, eCAADe, Liverpool, UK, pp. 169–177 (1999)

Bertol, D., Foell, D.: Designing Digital Space: An Architect's Guide to Virtual Reality. Wiley, New York (1997)

Bridges, A., Charitos, D.: On architectural design in virtual environments. Design Studies 18, 143–154 (1997)

Broadbent, G.: Design in architecture: architecture and the human sciences. John Wiley & Sons, London (1973)

Burry, M., Burry, J., Faulí, J.: Sagrada Família Rosassa: Global computeraided dialogue between designer and craftsperson (overcoming differences in age, time and distance). In: Reinventing the Discourse - How Digital Tools Help Bridge and Transform Research, Education and Practice in Architecture - Twenty First Conference of the Association for Computer Aided Design In Architecture, ACADIA, Buffalo, New York, USA, pp. 076–086 (2001)

Campbell, D. A., Wells, M.: A Critique of Virtual Reality in the Architectural Design Process, Seattle, USA (1994), http://www.hitl.washington.edu/publications/r-94-3/ accessed 2009/01/05

Chen, Y.-w.N.: Linking the virtual to reality: CAD & physical modeling. In: Tan, M., Teh, R. (eds.) The Global Design Studio - Sixth International Conference on Computer Aided Architectural Design Futures (CAADfutures), Centre for Advanced Studies in Architecture National University of Singapore, Singapore, pp. 303–311 (1995)

Dave, B.: Towards distributed computer-aided design environments. In: Tan, M., Teh, R. (eds.) The Global Design Studio - Sixth International Conference on Computer Aided Architectural Design Futures (CAADfutures), Centre for Advanced Studies in Architecture National University of Singapore, Singapore, pp. 659–666 (1995)

Davidson, J., Campbell, D.A.: Collaborative design in virtual space - Greenspace II: A shared environment for architectural design review. In: McIntosh, P., Ozel, F. (eds.) Design Computation: Collaboration, Reasoning, Pedagogy - Conference of the Association for Computer Aided Design In Architecture, ACADIA, University of Arizona, Tucson, Arizona, USA, pp. 165–179 (1996)

de Vries, B., Achten, H.H.: DDDoolz: designing with modular masses. Design Studies 23(6), 515–531 (2002)

Dobson, A.: Exploring conceptual design using CAD visualisation and virtual reality modelling. In: Computerised Craftsmanship - 16th Conference on Education in Computer Aided Architectural Design in Europe, eCAADe, Paris, France, pp. 68–71 (1998)

Dokonal, W., Hirschberg, U. (eds.): Digital Design, eCAADe and Graz University of Technology, Graz, Austria (2003)

Dorst, K.: The design problem and its structure. In: Cross, N., Christiaans, H., Dorst, K. (eds.) Analysing Design Activity, pp. 17–34. John Wiley & Sons Ltd., Chichester (1996)

Dorta, T.V.: L'influence de la réalité virtuelle non-immersive comme outil de visualisation sur le processus de design, Faculté des études supérieures, Université de Montréal, Montréal, Canada (2001)

Gero, J.S.: Where's the Research? The International Journal of Design Computing, IJDC 2(1) (1999a)

Gero, J.S.: Constructive memory in design thinking. Architectural Science Review 42, 3–5 (1999b)

Gibson, I., Kvan, T.: The use of rapid prototyping for architectural concept modelling, SME Technical Paper (PE02-222), p. 8 (2002)

Goel, V.: Sketches of Thought. MIT Press, Cambridge (1995)

Goldschmidt, G.: The dialectics of sketching. Creativity Research Journal 4, 123–143 (1991)

Gruber, A., Hirschberg, U., Dank, R.: Calculated Bananas: Defining a new introductory course in visual design for first year architecture students. In: Dokonal, W., Hirschberg, U. (eds.) Digital Design, 21st Conference on Education in Computer Aided Architectural Design in Europe, eCAADe and Graz University of Technology, Graz, Austria, pp. 519–522 (2003)

Hirschberg, U., Schmitt, G., Kurmann, D., Kolarevic, B., Johnson, B., Donath, D.: The 24 hour design cycle: An experiment in design collaboration over the internet. In: Fourth International Conference on Computer Aided Architectural Design Research in Asia, CAADRIA, Shanghai, China, pp. 181–190 (1999)

Kalay, Y.E.: P3: Computational environment to support design collaboration. Automation in Construction 8(1), 37–48 (1998)

Koutamanis, A.: Approaches to the integration of CAAD education in the electronic era: Two value systems. In: Brown, A., Knight, M.B.P. (eds.) Architectural Computing from Turing to 2000 - 17th Conference on Education in Computer Aided Architectural Design in Europe, eCAADe, Liverpool (UK), pp. 238–243 (1999)

Kurmann, D.: Sculptor - A tool for intuitive architectural design. In: Tan, M., Teh, R. (eds.) The Global Design Studio - Sixth International Conference on Computer Aided Architectural Design Futures (CAADfutures), Centre for Advanced Studies in Architecture National University of Singapore, Singapore, pp. 323–330 (1995)

Kvan, T.: Designing together apart, The Open University Milton Keynes England, Milton Keynes, UK (1999)

Kvan, T., Yip, W.H., Vera, A.: Supporting design studio learning: an investigation into design communication in computer-supported collaboration. In: CSCL 1999, Stanford University, pp. 328–332 (1999)

Lawson, B.: How designers think. Architectural Press, London (1980)

Lawson, B.: Design in mind. Butterworth-Heinemann Ltd., Oxford (1994)

Lawson, B., Loke, S.M.: Computers, words and pictures. Design Studies 18, 171–183 (1997)

Leach, N.: Designing for a digital world. Wiley, Chichester (2002)

Maher, M.L., Simoff, S.J., Cicognani, A.: Understanding Virtual Design Studios. Springer, London (2000)

Maver, T.: Three design paradigms: A tentative philosophy. DMG-DRS Journal 9, 130–132 (1973)

Maver, T.: Predicting the past, remembering the future. In: 6th IberoAmerican Congress of Digital Graphics, SIGraDi, Caracas, Venezuela, pp. 2–3 (2002)

Maze, J.: Virtual tactility: working to overcome perceptual and conceptual barriers in the digital design studio, thresholds - design, research, education and practice. In: The Space Between the Physical and the Virtual - 2002 Annual Conference of the Association for Computer Aided Design In Architecture, ACADIA, Pomona, California, USA, pp. 325–331 (2002)

McLuhan, M.: Understanding Media: The Extensions of Man. McGraw-Hill Book Company, New York (1964)

Mitchell, W.J.: Three paradigms for computer-aided design. Automation in Construction 3(2-3), 239–245 (1994)

Mitchell, W.J.: The future of the virtual design studio. In: Wojtowicz, J. (ed.) Virtual Design Studio, pp. 51–60. Hong Kong University Press, Hong Kong (1995)

Pérez-Gómez, A., Pelletier, L.: Architectural representation and the perspective hinge, Massachusetts Institute of Technology, Cambridge, Mass (1997)

Regenbrecht, H., Kruijff, E., Donath, D., Seichter, H., Beetz, J.: VRAM - A virtual reality aided modeller. In: Donath, D. (ed.) Promise and Reality: State of the Art versus State of Practice in Computing for the Design and Planning Process - 18th Conference on Education in Computer Aided Architectural Design in Europe, eCAADe, Weimar, Germany, pp. 235–237 (2000)

Robbins, E.: Why architects draw. MIT Press, Cambridge (1994)

Schmitt, G.: Shared authorship in design - Phase (X) and multiplying time, computers, Design Studio Teaching. In: EAAE/eCAADe International Workshop Proceedings, Leuven, Belgium, pp. 15–30 (1998)

Schnabel, M.A.: Collaborative studio in a virtual environment. In: Kinshuk, Lewis, R., Akahori, K., Kemp, R., Okamoto, T., Henderson, L., Lee, C.-H. (eds.), Learning Communities on the Internet - Pedagogy in Implementation, Proceedings of the International Conference on Computers in Education (ICCE), Auckland (New Zealand), pp. 337–341 (2002)

Schnabel, M.A.: Architectural Design in Virtual Environments, Ph.D. thesis, Department of Architecture, The University of Hong Kong, Hong Kong (2004)

Schnabel, M.A.: Framing mixed reality. In: Wang, X., Schnabel, M.A. (eds.) Mixed Reality Applications in Architecture, Design, and Construction, pp. 3–11. Springer, Netherlands (2009a)

Schnabel, M.A.: Interplay Of domains: New dimensions of design learning. In: Wang, X., Schnabel, M.A. (eds.) Mixed Reality Applications in Architecture, Design, and Construction, pp. 219–226. Springer, Netherlands (2009b)

Schnabel, M.A., Kvan, T.: Design, communication & collaboration in immersive virtual environments, International Journal of Design Computing, Special Issue on Designing Virtual Worlds, 4 (2002), http://faculty.arch.usyd.edu.au/kcdc/ijdc/vol04/papers/schnabelFrameset.htm (Last accessed: May 2009)

Schnabel, M.A., Kvan, T.: Spatial understanding in immersive virtual environments. International Journal of Architectural Computing (IJAC) 1(4), 435–448 (2003)

Schnabel, M.A., Wang, X., Seichter, H., Kvan, T.: From virtuality to reality and back. In: Proceedings of the 12th International Association of Societies of Design Research (IASDR), Hong Kong, pp. 12–15 (2007)

Schön, D.A., Wiggins, G.: Kinds of seeing and their functions in designing. Design Studies 13(2), 135–156 (1992)

Seichter, H.: VRAM/G - Gesture Enabled Virtual Reality Aided Modeller, Bauhaus University, Weimar, Germany (2001)

Stuart, R.: The Design of Virtual Environments. McGraw-Hill, New York (1996)

Sutherland, I.E.: Sketchpad, A Man-Machine Graphical Communication System. In: 1963 Spring Joint Computer Conference AFIPS, vol. 23 (1963)

Suwa, M., Tversky, B.: What do architects and students perceive in their design sketches? A protocol analysis. Design Studies 18, 385–403 (1997)

VDS: Virtual Design Studio, Hong Kong (1993-2002), http://courses.arch.hku.hk/vds (Last accessed May 2009)

Verstijnen, I.M., Hennessey, J.M., van Leeuwen, C., Hamel, R., Goldschmidt, G.: Sketching and creative discovery. Design Studies 19, 519–546 (1998)

Wenz, F., Hirschberg, U.: Phase(x) - Memetic engineering for architecture. In: Martens, B., Linzer, H., Voigt, A. (eds.) Challenges of the Future: 15th Conference on Education in Computer Aided Architectural Design in Europe, Österreichischer Kunst- und Kulturverlag, Vienna, Austria. (1997)

Wiener, N.: The Human Use of Human Beings: Cybernetics and Society, pp. 97–98. Houghton Mifflin, Boston (1954)

Wong, W., Kvan, T.: Textual support of collaborative design. In: Ataman, O., Bermúdez, J. (eds.) Media and design process - Conference of the Association for Computer Aided Design In Architecture, ACADIA, Salt Lake City, USA, pp. 168–176 (1999)

# Author Biographies

***Laura L Arns***, a Computer Scientist for the Weapons Lethality Analysis Branch of the NAWCWD, NAVAIR, China Lake, received her Ph.D. in Computer Science from Iowa State University in 2002. Prior to joining NAVAIR, she was the Associate Director and Research Scientist for the Envision Center at Purdue, and also held a courtesy Assistant Professor appointment with the Purdue Department of Computer Graphics Technology, where she taught a graduate course on virtual environments. Her research interests are in the areas of applied virtual environments, human factors and usability in virtual environments, and real-time graphics.

***Zeeshan Aziz*** is a lecturer at University of Salford, UK. His research interests include IT support for managing disasters involving critical physical infrastructure, application of novel technologies such as Remote Sensing, Virtual Reality and Geographic Information Systems in major civil engineering projects.

***Chiu-Shui Chan*** received his PhD in Architecture from Carnegie-Mellon University. He teaches a series of digital architecture courses at Iowa State University and is an affiliate professor of the Human Computer Interaction (HCI) Graduate Program at thc Virtual Reality Applications Center. He has served as visiting professor in Taiwan (1997, 2005) and China (2006). In 2006, he established a Digital Media Minor Program for the College of Design. He had been awarded National Science Foundation grants for developing the virtual model of the Inner City of Beijing. His recent book entitled: "Design Cognition: Cognitive Science in Design" was published by Chinese Architecture and Building Press (2008) in both English and Chinese versions. His areas of interest include: cognitive science, artificial intelligence in design, human computer interaction, virtual reality applications in architecture, and motion in design.

***Jian Chen*** received the PhD degree in computer science from Virginia Tech in 2006. She was a research associate in computer science at Brown University from 2006 to 2009. She is an assistant professor in the School of Computing at University of Southern Mississippi. Her research centers on 3D interaction, human-centered computing issues in scientific visualization, and computational modeling. She is a member of IEEE, ACM, and SIGCHI.

***Bharat Dave*** is an Associate Professor in the Faculty of Architecture, Building and Planning, University of Melbourne, Australia. He completed doctoral studies at the Swiss Federal Institute of Technology (ETH Zurich), and has held research and teaching positions in the USA, Switzerland, India, and Australia. His research revolves around innovative digital technologies in design at various scales, their cultural implications and emerging design futures. He initiated a research group CRIDA (Critical Research in Digital Architecture) in Melbourne, and currently supervises a number of research higher degree students. He is on the editorial board of the International Journal of Architectural Computing (UK), serves as an assessor for international research funding agencies, and regularly acts as a reviewer for a number of international journals and conferences. He was elected as President of the international association for the Computer Aided Architectural Design Research in Asia (CAADRIA) during 2005-08.

***Ellen Yi-Luen Do*** is currently an associate professor in the College of Architecture and the School of Interactive Computing at the College of Computing, affiliate faculty at the Health Systems Institute, and the Center for Music Technology at Georgia Institute of Technology where she directs ACME Lab and Healthcare Environment of the Future courses. Before joining Georgia Tech, she was on the faculty at Carnegie Mellon University (04-05) and on the faculty at University of Washington (99-04). Her work crosses the boundaries of design and human-computer interaction, cognitive science and studies of design. Her research explores new modalities of communication, collaboration, and coordination, as well as the physical and virtual worlds that push the current boundaries of computing environments for design and happy healthy living.

***Phillip S Dunston***, an Associate Professor with appointments in the Division of Construction Engineering and Management and the School of Civil Engineering at Purdue University, is a 2003 US National Science Foundation Career grantee for research on Mixed Reality applications for the architecture, engineering and construction (AEC) industry. He directs the Advanced Construction Systems Laboratory and is a Co-Director of the Consortium for Virtual Design of Healthcare Environments, both at Purdue.

***Andre Garcia*** is a graduate student in the Human Factors Applied Cognitive Psychology doctoral program at George Mason University. He earned his A.A. from Miami-Dade College and his B.A. from the University of Central Florida. His current research deals with visual-spatial cognition, 3D virtual environments, and human-computer interaction.

***Mani Golparvar-Fard*** is a Ph.D. Candidate in Construction Management and a Master of Computer Science student at the University of Illinois at Urbana-Champaign (UIUC). His research interests include automated visual sensing, semantic analysis and visualization of construction progress monitoring using daily construction photographs and 4D building information models.

***Ning Gu*** is a lecturer at the School of Architecture and Built Environment, University of Newcastle, Australia. He designed and implemented a wide variety of collaborative virtual environments and applied them in his research and teaching in numerous Australian and international tertiary design institutions including University of Newcastle, University of Sydney, MIT and Columbia University.

His research interests span the broad areas of architectural and design computing. He has published extensively in the field of architectural and design computing and design education. His career highlights include being a chief investigator in numerous research projects funded by the Australian Research Council (ARC) and the Australian Cooperative Research Centre for Construction Innovation (CRC-CI).

***Jeff WT Kan*** is a senior lecturer at the School of Architecture, Building and Design, Taylor's University College, Malaysia. He completed his PhD in design computing and cognition at the University of Sydney. During his study, he was awarded an International Postgraduate Research Award by the Australian Department of Education to undertake his PhD. His study focused on developing and using quantitative methods to study the cognitive behaviour of designers. He formerly taught design studio and computer-aided design at the Department of Architecture, Chinese University of Hong Kong. He has published papers on architectural visual information system, online interactive teaching materials, architectural visual impact studies, protocol analysis of designers, and methods to study design activities.

***Maria Kozhevnikov*** received her PhD in Technion – Israeli Institute of Technology. Since 2001, she has held faculty positions at the Harvard, Rutgers, and George Mason Universities. Her research interests focus on examining neural mechanisms of visual-spatial imagery as well as in exploring the ways to design visual-spatial training and assessments tools using three-dimensional immersive technologies. Currently, she is an Associate Professor of Psychology at the National University of Singapore and also, a Visiting Associate Professor at Harvard Medical School.

***Adam G Kushner***, a Lieutenant in the United States Navy and a former graduate student in the School of Civil Engineering at Purdue University, is currently an Instructor in the Naval Architecture and Ocean Engineering Department at the United States Naval Academy in Annapolis, MD.

***Gregory C Lasker***, an Assistant Professor in the Purdue University Department of Building Construction Management, is responsible for developing the Healthcare Construction Management specialty area. Prior to joining the program at Purdue, he amassed over twenty years of experience and leadership in residential, commercial, industrial and government construction, eleven of those as Owner/General Manager of Lasker Construction Group.

***Mary Lou Maher*** is a Program Director in the Information and Intelligent Systems Division at NSF and an Honorary Professor in the Design Lab at the University of Sydney. She joined the Human Centered Computing Cluster at NSF in July 2006 and started the CreativeIT program. Her research includes the development of cognitive and computational models of design and collaboration. She uses the protocol analysis method to demonstrate how new technologies impact design, collaboration, education, and creativity. She has studied the use of virtual worlds for education including MOOs, web-based learning environments, and 3D multi-user virtual worlds.

***James D Mcglothlin***, an Associate Professor of Industrial Hygiene and Ergonomics and the Director of the Occupational and Environmental Health Sciences Graduate Program in the Purdue University School of Health Sciences, is former Technical Director for the Regenstrief Center for Healthcare Engineering and current Co-Director of the Consortium for Virtual Design of Healthcare Environments, both at Purdue.

***Kathryn Merrick*** is a lecturer in information systems at the University of New South Wales, Australian Defence Force Academy. She completed her PhD in computer science through the National ICT Australia and the University of Sydney in 2007. She moved to ADFA in 2008. Her research interests span the broad fields of artificial intelligence, virtual worlds and developmental robotics. Specifically, she researches the design of computational models of motivation that allow machine learning algorithms to function in complex, dynamic environments. In such environments it may be difficult to predefine learning tasks or provide a teacher. Computational models of motivation use general task-independent concepts such as interest and curiosity to motivate the learning of task oriented behaviours. Applications of her research include the control of believable digital characters in online games, intelligent sensed environments, developmental robots and creative decision making.

***Rivka Oxman*** is Associate Professor and a Vice Dean at the Faculty of Architecture and Town Planning, Technion IIT. She is Associate Editor of the International Journal of Design Studies, Elsevier. She has received her doctoral degree from the Technion. She has held Visiting Professorship appointments at Stanford University USA; and Delft University of Technology Holland and research appointments at MIT and Berkeley U.S.A. Her current work explores the contribution of digital technologies to the emergence of new design paradigms. In 2002 she received the Design Research Society and Elsevier Science Award for the best paper of the year. In 2006 she was appointed as a Fellow of the DRS (Design Research Society). In 2007-08 she established and directed a program in digital design in the UK. Prof. Oxman has been appointed a member of the editorial boards of leading international scientific journals and conferences on design and computation and has published extensively in these and other journals.

***Feniosky Peña-Mora*** is Dean of Columbia University's Fu Foundation School of Engineering and Applied Science. He was previously Gutgsell Endowed Professor in the Department of Civil and Environmental Engineering and associate provost at the University of Illinois at Urbana-Champaign. He had been an associate professor at the Massachusetts Institute of Technology and visiting professor at international universities, including Ecole Polytechnique Fédérale de Lausanne. Dean Peña-Mora' research on construction engineering and management has been nationally-recognized, and he is the author of more than one-hundred publications in refereed journals, conference proceedings, book chapters, and textbooks on computer-supported design, computer-supported engineering design and construction, as well as project control and management of large-scale engineering systems.

***Aizhu Ren*** obtained her Doctoral degree (Structural Engineering) from Tsinghua University in 1992. She was the Director for Computer Application Division of Civil Engineering Department at Tsinghua University from July 1992 to March 2001, and the Director for Institute of Disaster Prevention and Reduction Engineering of Civil Engineering School since April 2001 to April 2007. She was an engineer of Building Design Institute, Tsinghua University from 1979 to 1985, and also was the member of Tsinghua Academic Committee, and the Chairman of the Academic Committee of Dept. of Civil Engineering and Dept. of Construction Management from 2004 to 2008. She served as a board member of International Society for Computing in Civil and Building Engineering (ISCCBE). She is currently the General Secretary of ISCCBE. She also serves as the Vice Chairman of National Society for Computing in Civil and Building Engineering of Chinese Civil Engineering Society.

***Seungjun Roh*** is a Ph.D. student in Construction Management at the University of Illinois at Urbana-Champaign (UIUC). His research interests include automation and visualization of interior construction progress monitoring using daily construction photographs and 4D as-planned models.

***Marc Aurel Schnabel*** is an Architect and Associate Professor at the *School of Architecture, The Chinese University of Hong Kong*. He is leading research and education in the field of digital media in architectural design. As President of CAADRIA, the international *Association for Computer Aided Architectural Design Research in Asia,* he is affiliated with various professional and scientific committees. He established the *Digital Architecture Research Alliance,* DARA, that brings together researchers who push the boundaries of current digital spatial design. He taught and worked in Germany Australia and Hong Kong for over fifteen years and since then has become highly recognised for his work in the areas of Virtual Environments and Design Learning. He publishes extensively in international journals about novel perspectives in Digital Architecture and the communication of three-dimensional space using innovative design methods and recently curated two Digital Architectural exhibitions, *Disparallel Spaces* at the *Tin Sheds Gallery* and *8448 cubed* at *Gaffa Gallery.*

***Fangqin Tang*** obtained his Doctoral degree (Disaster Prevention and Mitigation Engineering) from Tsinghua University in 2009. His research mainly focuses on disaster prevention and associated computational methods. He investigated the improved methods for quick modeling of irregular structures and applied it in the 3d modeling of Beijing National Olympic Stadium - Bird's Nest (2005). He was one of the main participants of the Joint project with Hong Kong Polytechnic University - "Integration of GIS, CAD and Virtual Reality for Construction" (2006) and also involved in two supporting subjects of disaster research for National 11th 5 year Science and Technology Development Plan of China from 2007. He studied and developed a 3d city navigation system with the expertise on computer graphics and 3d rendering during the Intel graduate technical intern (2008).

***Walid Tizani*** is an associate professor in the Department of Civil Engineering, the Faculty of Engineering, the University of Nottingham, UK. His main research areas are the application of novel information technologies to engineering design and the investigation of structural performance of steel frames and connections. His experience includes: the application of virtual prototyping, virtual reality techniques, knowledge based systems and object oriented technology in the areas of engineering design, decision support systems; construction-led design, and integrated and collaborative design. He is a member of the Board of Directors of the International Society for Civil and Building Engineering and the chair of the ICCCBE 2010 conference. He is currently a Vice-Dean of the Faculty of Engineering and lectures in Steel Structures, IT for Engineers, and Engineering Communication.

***Jerry J-H Tsai*** is a researcher in design computing. He was awarded his PhD from the University of Sydney. His research interests include design representation, system integration, and collaborative design. He is an architect and interior designer focusing on residential design. He has been a lecturer in Australia and Taiwan coordinating and teaching design studios, creative design, collaborative virtual environments and computer graphic concepts. Currently, he is an assistant professor at Yuan Ze University Taiwan, working on different collaborative projects with researchers in Australia, Malaysia, Singapore and Taiwan. He is also a co-director of C-Lab, which is a cross-discipline research and design lab at Yuan Ze University. It involves people from Art & Design department and Mechanical Engineering department. He works on human-computer interaction, ambient intelligence, and smart environments projects at the lab.

***Xiangyu Wang*** is a Senior Lecturer in the Faculty of Built Environment at The University of New South Wales, Australia. He obtained his Ph.D. degree in Civil Engineering at Purdue University in 2005. Dr. Wang's work is featured with highly interdisciplinary research across Design, Computer Engineering, Construction, and Human Factors. His specific research interests include virtual environments for design, human-computer interactions, computer-supported cooperative work, and construction automation and robotics. He is now supervising five Ph.D. students and has published over 140 refereed articles into a wide range of highly recognized international journals and conferences (ASCE, IEEE, ACM, etc.). He was also awarded a US National Science Foundation grant to investigate skill development through virtual technologies.

***Rui Wang*** is a Master by Philosophy student in Design Lab, Faculty of Architecture, Design and Planning, the University of Sydney. Her research interests are presence issues in Mixed-Reality supported systems and environments under the supervision of Dr. Xiangyu Wang.

***Nobuyoshi Yabuki*** is a Professor in the Division of Sustainable Energy and Environmental Engineering, Graduate School of Engineering, Osaka University, Japan. He leads the Environmental Design and Information Technology Laboratory there. He obtained his Ph.D. and M.S. at Stanford University and B.E. at the University of Tokyo. His research interests include the development of product models, Building Information Modeling (BIM), application of VR and AR to civil and building engineering, 4D CAD, RFID and sensors for inspection and monitoring of structures, data mining and knowledge discovery from large amount of sensor and product model data, cooperative design, IT in construction. He leads Civil Engineering Committee of buildingSMART IAI Japan and Cyber & Real Infrastructure Modeling Committee of Japan Society of Civil Engineers (JSCE).

# Author Index

# Index